AF294878

"The shoulders on which we stand"

125 Jahre Technische Universität Berlin

Springer-Verlag

Berlin Heidelberg GmbH

Veröffentlicht mit freundlicher Unterstützung von
Published with friendly support of

"The shoulders on which we stand" – Wegbereiter der Wissenschaft

125 Jahre Technische Universität Berlin

Im Auftrag des Präsidenten
der Technischen Universität Berlin

Herausgegeben von Eberhard Knobloch

 Springer

Technische Universität Berlin
Straße des 17. Juni 135
10623 Berlin

ISBN 978-3-642-62353-0   ISBN 978-3-642-18916-6 (eBook)
DOI 10.1007/978-3-642-18916-6

Bibliografische Information der Deutschen Bibliothek
Die Deutsche Bibliothek verzeichnet diese Publikation in der Deutschen Nationalbibliografie; detaillierte
bibliografische Daten sind im Internet über <http://dnb.ddb.de> abrufbar.

Einbandentwurf: deblik Berlin
Satz: medio Technologies AG, Berlin
Gedruckt auf säurefreiem Papier   68/3020/M-5 4 3 2 1 0

Vorwort des Herausgebers

Die bewegte Geschichte der Technischen Universität Berlin (TUB) reicht durch deren ältestes Lehr- und Forschungsgebiet „Bergbau und Geowissenschaften" bis in die Zeit König Friedrich II. zurück. Die 1770 gegründete Bergakademie wurde 1916 der Königlichen technischen Hochschule zu Berlin angegliedert, die selbst 1879 durch Zusammenschluss der 1799 gegründeten Bauakademie und der 1821 gegründeten Gewerbeakademie entstanden war.

Aber auch später erweiterte die TUB ihre Lehr- und Forschungsaufgaben durch Eingliederung von Teilen anderer Hochschulen. 1927 übernahm sie das geodätische Institut der Landwirtschaftlichen Hochschule, 1980 Fachgebiete der Pädagogischen Hochschule. Ihre Entstehungsgeschichte bringt es mit sich, dass zu verschiedensten Zeiten Jubiläen der Vorgängerinstitutionen gewürdigt wurden.

Die vorliegende Festschrift nimmt das Gründungsjahr 1879 der Technischen Hochschule als Ausgangspunkt. Da 1979 anlässlich des hundertjährigen Jubiläums eine zweibändige, auf historischen Beiträgen beruhende Festschrift (Herausgeber Reinhard Rürup) erschien, schien es sinnvoll, 25 Jahre später einen anderen Zugang zu wählen. Der wissenschaftliche Ruf einer Universität hängt entscheidend von deren Hochschullehrern ab. Deshalb entschied die Universität, fünfundfünfzig herausragende Persönlichkeiten ihrer Professorenschaft bzw. Forscher durch Kurzbiographien vorzustellen und so eine Art Kaleidoskop ihrer Lehr- und Forschungstätigkeit zu veröffentlichen. Außer wissenschaftlicher Exzellenz sollten die aufzunehmenden Personen wenigstens fünf Jahre im Ruhestand sein, um das erforderliche Maß an objektiver Beurteilung sicherzustellen.

Die acht Fakultäten wurden um Vorschläge gebeten, um die Fachgebiete ausgewogen und angemessen zu repräsentieren. Die Entscheidung über die Auswahl lag bei den beiden früheren TU-Präsidenten Manfred Fricke, Jürgen Starnick und mir. Wir sind sicher, neben den drei Nobelpreisträgern Gustav Hertz, Ernst Ruska und Eugene Wigner würdige Kandidaten ausgewählt zu haben. Dies schließt nicht aus, dass im Einzelfall

Vorwort des Herausgebers

ein anderes Auswahlgremium vielleicht gelegentlich andere Persönlichkeiten vorgezogen hätte.

Die Biographien wurden von den vier Wissenschaftshistorikern Brita und Michael Engel (B.E., M.E.), Frank Horstmann (F.H.) und Jörg Zaun (J.Z.) erarbeitet und von Cormac Deane und Robert Sleigh ins Englische übertragen. Die vertrauensvolle Zusammenarbeit mit Ursula Weiß, Hans-Jürgen Süssespeck vom Präsidialamt der TU und Thomas Lehnert vom Springer-Verlag sei dankend hervorgehoben. Bei der Besorgung der Bildvorlagen hat sich Romy Werther verdient gemacht.

Berlin, im Februar 2004   *Eberhard Knobloch*

The eventful history of the Technische Universität Berlin (TU) can be traced back to the reign of Friedrich II, when Mining and Earth Sciences, the university's oldest field of learning and research, was first studied. The Bergakademie (Mining Academy), which was founded in 1770, was incorporated into the Königliche Technische Hochschule in 1916, which itself was the product of a merger in 1879 between the Bauakademie (founded 1799) and the Gewerbeakademie (founded 1821).

The TU continued to expand its fields of teaching and research by incorporating departments of other institutions of higher learning. In 1927, the university took over the Geodätisches Institut der Landwirtschaftlichen Hochschule and in 1980 it started teaching courses of the Pädagogische Hochschule. The nature of the TU's formation over the years means that it has a regular schedule of the anniversaries of its original institutions to celebrate.

This jubilee volume takes the year 1879, when the Technische Hochschule was founded, as its starting point. A two-volume publication that recounted the history of the institution was published in 1979 under the editorship of Reinhard Rürup, and now, 25 years on, we are taking a different approach. The scientific reputation of a university is based to a great extent on its teaching staff. That is why the university has decided to use this volume to present short biographies of 55 outstanding personalities from among its professors and researchers, thereby displaying the full spectrum of the TU's achievements as an institution of teaching and research. The selection of subjects was made according to two criteria: each individual demonstrated excellence in science, and each has been in retirement for at least five years, in order that a proper degree of objectivity can be achieved.

The eight faculties of the TU were asked to suggest individuals so that the breadth of the university's fields of study would be represented in a measured and balanced way. The decision on who should be included was made by two former TU presidents, Manfred Fricke and Jürgen Starnick, and by me. We feel sure that, alongside the three Nobel prizewinners

Gustav Hertz, Ernst Ruska and Eugene Wigner, we have chosen worthy candidates. This of course does not exclude the possibility that another selection panel may have chosen a different list of names.

The biographies in this volume were written by four historians of science: Brita Engel (B.E.), Michael Engel (M.E.), Frank Horstmann (F.H.) and Jörg Zaun (J.Z.). They were translated into English by Cormac Deane and Robert Sleigh. For their reliable assistance throughout, I would like to thank Ursula Weiß and Hans-Jürgen Süssespeck from the Präsidialamt and Thomas Lehnert from Springer-Verlag Publishers. Romy Werther was indispensable in sourcing picture material for this volume.

Berlin, February 2004   *Eberhard Knobloch*

Grußwort des Präsidenten
der Technischen Universität Berlin

Liebe Leserinnen und Leser,

vor 125 Jahren entstand durch die Vereinigung von Bau- und Gewerbeakademie die „Königliche technische Hochschule zu Berlin". Die Technische Universität Berlin nimmt dies zum Anlass, 2004 das Jubiläum dieser Vorläufereinrichtung zu begehen. Die 1879 neu gegründete Hochschule konnte die Wurzeln ihrer Teilinstitutionen allerdings noch weiter zurückverfolgen: so feierte die eben erst aus der Taufe gehobene Lehr- und Forschungsanstalt bereits 1899 – also nur zwanzig Jahre später – ihr 100jähriges Jubiläum, in Berufung auf die Gründung ihrer Vorläufereinrichtung, der Bauakademie 1799. Waren die Jubiläumsfeierlichkeiten am Ende des 19. Jahrhunderts Ausdruck ungebrochenen Selbstbewusstseins und glanzvolle Inszenierung der Technikelite, hat sich das Selbstverständnis der TU Berlin heute grundlegend gewandelt. Mit der Wiedergründung der TH Berlin als „Technische Universität Berlin" durch die Briten im Jahr 1946 ist eine weltoffener, den ethischen Grundsätzen von Forschung und Lehre verpflichteter Ort entstanden, der sich der Verantwortung seiner Geschichte, auch gegenüber deren dunklen Kapiteln, bewusst ist.

Universität ist ein Raum, der erst durch die Persönlichkeiten, die an ihr lehren und forschen, seine Vitalität und seinen wissenschaftlichen Ruf erhält. Aus diesem Grund haben wir uns entschieden, statt einer Institutionsgeschichte die fünf Vierteljahrhunderte unseres Bestehens in Form von Lebensläufen von Wissenschaftlern darzustellen, die unsere Einrichtung geprägt haben. Ihre Arbeit hinterließ nicht nur Spuren im Profil unserer Universität, sie hat auch dem wissenschaftlichen Fortschritt den Weg bereitet. Ohne den Erkenntnisgewinn vorausgegangener Generationen von Wissenschaftlerinnen und Wissenschaftlern, ist wissenschaftliche, wirtschaftliche und schließlich auch gesellschaftliche Weiterentwicklung nicht möglich. Es fiel uns daher leicht, dies schon – in Anlehnung an Newton – im Titel zu formulieren: „The shoulders on which we stand – Wegbereiter der Wissenschaft".

Grußwort des Präsidenten
der Technischen Universität Berlin

55 Wegbereiter, die seit 1879 an der TU Berlin bzw. ihren Vorläufereinrichtungen gelehrt und geforscht haben, werden in dem vorliegenden Band vorgestellt. Die schwierige Auswahl hat die kompetente Jury, bestehend aus Prof. Dr. Eberhard Knobloch sowie den Altpräsidenten Prof. Dr.-Ing. Manfred Fricke und Prof. Dr. Jürgen Starnick, getroffen, denen ich dafür zu großem Dank verpflichtet bin. Herr Prof. Knobloch hat darüber hinaus die Arbeit als Herausgeber übernommen, wofür ich ihm ebenfalls herzlich danke. Das Unternehmen hätte keine Aussicht auf Erfolg gehabt, hätten sich nicht vier Wissenschaftshistoriker gefunden, die in eindrücklichen Texten die Protagonisten unserer Publikation gleichsam „zum Leben erweckt" haben: mein Dank geht an Frau Brita Engel, Herrn Dr. Michael Engel, Herrn Dr. des. Frank Horstmann und Herrn Dr. Jörg Zaun. Ich freue mich sehr, dass der Springer-Verlag – wie schon für frühere Festschriften – auch für dieses Projekt gewonnen werden konnte. Die Zusammenarbeit, insbesondere in Person von Herrn Thomas Lehnert, gestaltete sich ebenso effektiv wie angenehm.

Ich darf Ihnen nun 55 interessante Entdeckungsreisen durch Leben großer Forscher und Lehrer wünschen, viel Vergnügen bei dem Streifzug durch 125 Jahre Wissenschaftsgeschichte.

Im Februar 2004   *Prof. Dr. Kurt Kutzler*

Message from the President
of the Technical University of Berlin

Dear reader,

It is 125 years since the Bauakademie and the Gewerbeakademie merged to become the Königliche technische Hochschule zu Berlin, and it is this anniversary that the Technical University (TU) has chosen to celebrate in 2004. The Technische Hochschule, which was founded in 1879, was able to trace the roots of its constituent parts even further into the past. Thus in 1899, only twenty years after it was founded, the TH celebrated the 100th anniversary of the foundation of the Bauakademie. While the jubilee celebrations of the tail end of the nineteenth century were an expression of the undiminished self-confidence and glamorous self-image of the technical elite, the TU's sense of itself is fundamentally different in the present day. When the TH Berlin was transformed into the Technische Universität of Berlin by the British occupation forces in 1946, an institution was founded that was open to the world at large and dedicated to the ethical principles of research and education, and which is conscious of the responsibilites it bears for its history, including its darker chapters.

A university is a place whose vitality and academic reputation are formed on the basis of the people who teach and research there. For this reason we have decided, rather than publish an institutional history, we would reflect the achievements of more than a century of excellence in the careers of the scientists who have left their mark on the university. Their work has not only distinguished the TU itself, but it has also paved the way for important advances across the sciences. Economic, social and scientific development are only possible if built on the foundations of knowledge set down by previous generations of scientists. Hence the title of the current volume, which is derived from Isaac Newton's maxim "If I have seen farther, it is by standing on the shoulders of giants."

This book presents 55 ground-breaking scientists who taught and researched in the TU and in its predecessor institutions. The difficult task of deciding whom to include fell to the competent jury made up of Professor Eberhard Knobloch and former TU presidents Professor Manfred

Fricke and Professor Jürgen Starnick, and I am very grateful to them all for their work. Professor Knobloch also undertook to edit the volume, and I thank him for that. The whole undertaking would have had no prospect of success without the contributions of four science historians; Brita Engel, Dr. Michael Engel, Dr. Frank Horstmann and Dr. Jörg Zaun. Their texts brought to life the careers of the protagonists of this volume. I was delighted that we were able once again to convince Springer Verlag publishers, who have helped us with previous commemorative publications, to join us in the current project. It was a pleasure to work with them, in particular with Thomas Lehnert.

It remains for me only to wish you an enjoyable journey of discovery through the lives of 55 great researchers and teachers and through 125 years of the history of science in the pages ahead.

February 2004   *Professor Kurt Kutzler*

Inhalt/Table of Contents

Inhalt/Table of Contents

125 Jahre Technische Universität Berlin

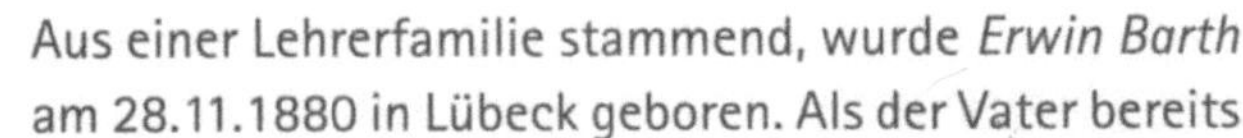

Aus einer Lehrerfamilie stammend, wurde *Erwin Barth* am 28.11.1880 in Lübeck geboren. Als der Vater bereits 1883 starb, gelang es der Mutter, die damit entstehenden wirtschaftlichen Schwierigkeiten zu überbrücken. Um Mutter und Schwester zu entlasten, verließ der naturwissenschaftlich hochbegabte und interessierte Barth mit der Obersekundareife die Schule und begann eine Gärtnerlehre. Nach dreijähriger Lehr- und Gehilfenzeit ging er an die Kgl. Preuß. Gärtner-Lehranstalt in Wildpark, wo er sich besonders der bildenden Gartenkunst widmete.

Als Gartentechniker begann Barth 1902 in Hannover seine Laufbahn, wurde 1903 nach Bremen empfohlen, arbeitete 1904 als „erster Gartenarchitekt" bei einer Düsseldorfer Firma, war zusätzlich mehrfach in öffentlichem und privatem Auftrag in Hannover tätig, leistete 1903/04 auch den Militärdienst, bis 1906 an der Gärtnerlehranstalt – inzwischen nach Dahlem verlegt – die Obergärtner-Prüfung in der Gartenkunst erfolgte. Seit 1909 konnte er sich „Staatlich diplomierter Gartenmeister" nennen. 1906–08 arbeitete er als Geschäftsführer in Köln, entwarf eine Reihe privater Hausgärten und nahm erfolgreich an einigen Wettbewerben teil. 1908 erfolgte die Berufung zum Stadtgärtner in Lübeck. Hier fand er Anlass, sich ausgiebig mit Problemen des Friedhofswesens, der Spiel- und Sportanlagen, der Kleingärten und des Natur- und Landschaftsschutzes auseinanderzusetzen. Weiterhin entwickelte er seine Grundsätze der Landschaftsgestaltung und naturbezogener Landschaftsplanung. Barth erweist sich als Mittler zwischen dem herkömmlichen Stil der Architekturgärten und den modernen, ökologisches Gedankengut aufnehmenden Konzepten. Die seiner Ausbildungsstätte gemäße Bewahrung und Weiterentwicklung Lennéscher Vorstellungen ist auch bei ihm unverkennbar, sofern der originale Zustand seiner Garten- und Parkanlagen überhaupt noch rekonstruierbar oder gar erkennbar und damit interpretierbar ist.

Barth liebte seine Heimatstadt, war aber mit seinen dienstlichen Kompetenzen höchst unzufrieden, weshalb er sich auf freie Stel-

Erwin Barth, whose parents were teachers, was born in Lübeck on 28 November 1880. When the father died only three years later, his mother managed to overcome the ensuing economic hardships. Barth, who was highly gifted and interested in science, left school at an early stage and began an apprenticeship as a gardener to ease the burden on his mother and his sister. After three years as an apprentice and assistant, he attended the Royal Prussian School for Gardeners at Wildpark, where he showed particular interest in landscape gardening.

In 1902, Barth began his career as a garden technician in Hanover and in 1903, followed a recommendation that took him to Bremen, before he began working as a "senior landscape gardener" for a company in Dusseldorf. At the same time, he was also busy with public and private contracts in Hanover, and performed his military service in 1903/04, before returning to the Garden School (which had meanwhile moved to the Dahlem district of Berlin) in 1906, to take his senior gardener's examination in landscape gardening. From 1909 onwards, he was allowed to call himself a *Staatlich diplomierter Gartenmeister* (State Examined Senior Gardener). He held a managerial post in Cologne from 1906 to 1908, designing a series of gardens for private houses and successfully taking part in a number of competitions. In 1908 he was offered a position as the city of Lübeck's gardener in chief. It was here that he was able to give much thought to the subjects of cemeteries, sports fields, playgrounds, and allotments, as well as the problems of nature conservation and conservation areas. In addition, he developed his own basic principles of landscape gardening and conservation-minded landscape planning. Barth proved to be a mediator between the conventional style of landscaping and modern, ecological-influenced ideas. What is also unmistakable in those places where it has been possible to restore his gardens and parks to their original state (or even to get an idea of how they looked) is his adherence to and further development of the ideas of

len bewarb, so auch auf die des Gartendirektors von Charlottenburg, mit dem Ziel, durch Bleibeverhandlungen bessere Bedingungen zu erreichen. Der Lübecker Senat blieb starr, doch um das Gesicht zu wahren, ging er 1912 nolens volens nach Charlottenburg. Diese damals sehr wohlhabende Stadt ermöglichte ihm, seine Vorstellungen in künstlerisch-ästhetischer wie auch in sozialer Hinsicht (Volksparkgedanke) zu verwirklichen. Dabei war er so erfolgreich, daß er 1925 als einziger Kandidat aufgestellt und zum Stadtgartendirektor Groß-Berlins gewählt wurde. Seit 1921 an der TH Dozent, seit 1927 Honorarprofessor für Gartenkunst, setzte er sich für die wissenschaftliche Aufwertung der Gartenarchitektur und -gestaltung zum Hochschulfach ein. 1929 wurde er zum Ordinarius für Gartengestaltung an die Landwirtschaftliche Hochschule Berlin berufen und war damit seinem hochschulpolitischen Ziel ein wesentliches Stück näher gekommen. Schon Anfang 1933 geriet er in Konflikt mit den Nationalsozialisten und befürchtete die Einweisung in ein Konzentrationslager. Barth starb am 10. Juli 1933 durch Selbstmord.

Auf seine Entwürfe gehen u.a. zurück: Lietzenseepark, Brixplatz (1921 als „ökologische" Anlage fertiggestellt), Karolingerplatz, Mierendorffplatz, Volkspark Jungfernheide und der Wilmersdorfer Waldfriedhof in Stahnsdorf.

Lit.: Erwin Barth. Gärten, Parks, Friedhöfe; Katalog zur Ausstellung, Berlin 1980.

[M. E.]

the German landscape gardener Peter Joseph Lenné, in accordance with the principles of the college he had attended. Barth loved his home town, but he was absolutely discontented with his own official powers, which is why he applied for other jobs (such as director of gardens in Charlottenburg, Berlin) in the hope that, in time, he would be able to negotiate better conditions. The city council in Lübeck remained inflexible, so in 1912, in order to save his face, he left for Charlottenburg, willingly or unwillingly. Charlottenburg, then a very affluent city, let him put his ideas into practice – both in the artistic and aesthetic senses, as well as in the social sense (i.e. the idea of a people's park). He was so successful that he was nominated as the only candidate and elected the Greater Berlin city garden director in 1925. Barth, a lecturer at the TH since 1921, and from 1927 an honorary professor for landscape gardening, became committed to enhancing the status of garden design and landscaping so that it could become one of the subjects taught at university. In 1929 he was appointed professor for garden design at the Landwirtschaftliche Hochschule (Agricultural University) in Berlin and had thus come a great deal closer to reaching his academic goal. In early 1933, he came into conflict with the National Socialists and, fearing deportation to a concentration camp, took his own life on 10 July 1933.

The following are some of his designs that were realised in Berlin: Lietzenseepark, Brixplatz (completed as an "ecological area" in 1921), Karolingerplatz, Mierendorffplatz, the Volkspark Jungfernheide and the Wilmersdorf Waldfriedhof (forest cemetery) in Stahnsdorf.

Lit.: Erwin Barth. Gärten, Parks, Friedhöfe; Katalog zur Ausstellung. Berlin 1980.

[M. E.]

Hans Boersch (1909–1986)

Hans Boersch wurde am 1. Juni 1909 in Berlin geboren. Er begann sein Physikstudium 1930 an der Technischen Hochschule Charlottenburg. 1933 wechselte er an die Universität Wien, wo er 1935 mit einer Arbeit „Bestimmung der Struktur einiger einfacher Moleküle mit Elektronen-Interferenz" promovierte. Noch im gleichen Jahr wurde Boersch wissenschaftlicher Mitarbeiter im Forschungslaboratorium der AEG in Berlin. Hier untersuchte er die elektronenoptische Bildentstehung erstmals unter wellenoptischen Gesichtspunkten. Er entwickelte das Elektronen-Schattenmikroskop, mit dem ihm 1939 der Nachweis der Beugung von Elektronen an makroskopischen Kanten gelang und damit der endgültige experimentelle Nachweis der Wellennatur von Elektronen. 1941 erhielt er zusammen mit mehreren anderen Pionieren der Elektronenmikroskopie die Silberne Leibniz-Medaille der Preußischen Akademie der Wissenschaften verliehen.

Boersch kehrte 1941 an die Universität Wien zurück, wo er als Assistent am I. Chemischen Institut das Laboratorium für Strukturforschung und Übermikroskopie aufbaute. Er habilitierte sich 1942 mit einer Arbeit über Elektronen-Beugungserscheinungen und er entwickelte das Ionen-Übermikroskop. Nach einer kurzen Zwischenstation als Privatdozent an der Universität Innsbruck wurde Boersch 1946 Mitarbeiter des Institut des Recherches Scientifiques in Tettnang (Württemberg). Auch hier baute er ein Laboratorium für Elektronenoptik auf und arbeitete weiter an der Entwicklung der Ionenmikroskopie.

1948 wechselte Boersch zur Physikalisch-Technischen Bundesanstalt in Braunschweig und erhielt gleichzeitig eine Honorarprofessur an der TH Braunschweig. Hier entdeckte er die anomale Energieverteilung in intensiven Elektronenstrahlen - den „Boersch-Effekt" –, die als begrenzender Faktor von entscheidender Bedeutung für das Auflösungsvermögen von Elektronenmikroskopen ist.

Hans Boersch was born in Berlin on 1 June 1909. He started studying physics at the Technische Hochschule Berlin in 1930, before enrolling at the University of Vienna in 1933, where he took a doctorate in 1935 with a paper on the specification of the structure of several simple molecules by means of electron interference. In the same year, Boersch took a post in the research laboratories of AEG in Berlin, where he investigated electron-optical image formation from the point of view of wave optics for the first time. He developed the shadow microscope, which he used in 1939 to prove the diffraction of electrons on macroscopic edges, thereby providing conclusive experimental proof of the wave-like nature of electrons. In 1941, he was among a group of pioneers who were awarded with the silver Leibniz medal from the Prussian Academy of Sciences for their work on electron microscopy.

Boersch returned to the University of Vienna in 1941, where as an assistant at the Chemistry Institute he built a laboratory devoted to research into atomic structure and electron microscopy. In 1942 he earned a professorship with a paper on diffraction phenomena and he developed the ion/electron microscope. After a short period as a private tutor at the University of Innsbruck, Austria, Boersch joined the Institut des Recherches Scientifiques in Tettnang (Württemberg) in 1946. There he also built a laboratory for electron optics and he continued his work on ion microscopy.

In 1948, Boersch started work at the Physikalisch-Technische Bundesanstalt (National Institute of Natural and Engineering Science) in Braunschweig, becoming at the same time honorary professor at the TH Braunschweig. During this period he discovered the phenomenon of anomalous energy distribution in intense electron beams, now known as the Boersch-effect, which is of central importance as a limiting factor for the resolution capability of electron microscopes.

1954 übernahm Boersch die Leitung des I. Physikalischen Instituts der TU Berlin. In Berlin entstanden Pionierarbeiten zur Holografie, zur Elektronenmikroskopie bei tiefen Temperaturen mit heliumgekühlten Objekten, zu supraleitenden magnetischen Linsen und zur Untersuchung ferromagnetischer Objekte unter dem Elektronenmikroskop, die in Verbindung mit der Tiefsttemperaturtechnik schließlich auch zur Beobachtung einzelner magnetischer Flussquanten in Supraleitern führte.

Als Anfang der 60er Jahre der Laser entwickelt wurde, begann Boersch umgehend mit dem Aufbau einer Forschungsgruppe an seinem Institut. Bald konnte er auch auf diesem Gebiet Erfolge vorweisen, so die Entwicklung eines hochkonstanten Lasers, die Erzeugung sehr kurzer Laserpulse oder die Entwicklung von Ionenlasern hoher Leistung.

Von 1957 bis 1959 war Boersch Dekan der Fakultät Allgemeine Ingenieurwissenschaften der TU. In den Jahren 1963 bis 1965 war er Vorsitzender der Deutschen Gesellschaft für Elektronenmikroskopie, die ihn 1973 zum Ehrenmitglied wählte. 1974 wurde Boersch emeritiert; für seine Verdienste um die Physik an der TU wurde er 1985 zum Ehrenmitglied der Technischen Universität ernannt. Boersch verstarb am 9. Juni 1986 nach kurzer schwerer Krankheit in Eggenfelden in Niederbayern.

Lit.: H. Niedrig: Nachruf auf den Pionier der Elektronenmikroskopie: Hans Boersch 1909–1986, Optik 75 (1987), S. 172–174

[J. Z.]

In 1954, Boersch became the director of the 1st Physics Institute at the TU Berlin. While at the TU, he made pioneering work into holography, electron microscopy at low temperatures with helium-cooled objects, superconducting magnetic lenses, and the investigation of ferromagnetic objects under the electron microscope, which, in conjunction with super-low-temperature technology, led finally to the observation of magnetic flux quantums in superconductors.

As the laser was being developed at the start of the 1960s, Boersch promptly set up a research group in his institute. And soon enough he was able to present successful work in the field, such as the development of high intensity lasers, very short laser pulses, and the development of high performance ion lasers.

From 1957 to 1959 Boersch was dean of the faculty of General Engineering Sciences at the TU. He served as the chairman of the German Society of Electron Microscopy from 1963 to 1965, which made him an honorary member in 1973. He became professor emeritus in 1974. He was made an honorary member of the TU in 1985 for his services to physics at the university. Boersch died on 9 July 1986 in Eggenfelden in Bavaria after a short illness.

Lit.: H. Niedrig: Nachruf auf den Pionier der Elektronenmikroskopie: Hans Boersch 1909–1986, Optik 75 (1987), p. 172–174

[J. .Z.]

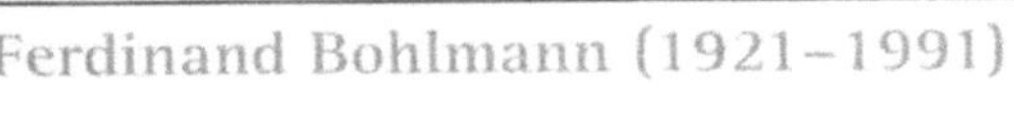

Ferdinand Bohlmann wurde am 28.8.1921 in Oldenburg geboren und begann im Kriegsjahr 1939 in Göttingen das Chemiestudium. Trotz Kriegsdienst und Verwundung erwarb er bereits 1944 das Diplom und wurde mit einer bei Hans Brockmann (1903–1988) angefertigten Dissertation über Solvatochromie in der Pyridinreihe 1946 in Göttingen promoviert. Anschließend arbeitete in Marburg bei Hans Herloff Inhoffen (1906–1992) am physiologisch-chemischen Institut, dem er nach Braunschweig folgte. Nach der Habilitation wurde er 1952 Diätendozent und 1957 apl. Professor. Bereits 1954 wurde er mit dem wenige Jahre vorher eingerichteten Dozentenpreis des Fonds der Chemischen Industrie ausgezeichnet, 1958 erfolgte die Verleihung des Göttinger Akademiepreises. Für das nach dem Weggang von Friedrich Weygand (1911–1969) freigewordene Ordinariat für organische Chemie war Bohlmann für die TU der geeignete Nachfolger. 1959 nahm er die Arbeit am Institut für Organische Chemie auf, und bald wuchs die Mitarbeitergruppe zu stattlicher Größe. 1974 wurde er von der GdCh mit der Verleihung der Otto-Wallach-Plakette geehrt. Trotz vieler internationaler Kontakte und weitreichender Anerkennung seiner Leistung blieb es ausschließlich bei diesen nationalen Würdigungen. Er starb am 23. September 1991.

Bohlmanns Leidenschaft galt den Naturstoffen, ihrer Isolierung, Strukturaufklärung und Synthese. Er war ein Meister im geschickten Einsatz der spektroskopischen Methoden „Bohlmann-Bande" in der IR-Spektroskopie, in der Entwicklung der synthetischen Verfahren und dem effizienten Einsatz der Mitarbeiter. Ausgezeichnet durch ein bewundernswertes Gedächtnis und getrieben von großer Ungeduld, ohne dabei in der Präzision Abstriche zu machen, machten Disziplin und Beharrlichkeit ihn mit seinem Arbeitskreis zu einem der weltweit produktivsten Teams. 1453 Veröffentlichungen sind das sichtbare Ergebnis.

Sein Hauptinteresse galt den Terpenen und den Polyinen, und es ließ sein Herz höher schlagen, als er diese auch in der Natur fand. Noch

Ferdinand Bohlmann was born on 28 August 1921 in Oldenburg and started studying chemistry at Göttingen in 1939, the year World War II broke out. Despite serving in the military and being wounded, he took his degree in 1944 and two years later he received a doctorate at Göttingen for a dissertation completed with Hans Brockmann (1903–1988) on solvatochromic phenomena in the pyridine series. Following that, he worked at the Physiological-Chemical Institute in Marburg with Hans Herloff Inhoffen (1906–1992), whom he followed to Braunschweig. He made the grade of professor in 1952, upon which he became a lecturer until he took a professorship in 1957. As early as 1954 he was awarded the recently-instituted lecturer's prize of the Fonds der Chemischen Industrie, and in 1958 he received the prize of the Göttingen Academy. When the departure of Friedrich Weygand (1911–1969) left the chair of Organic Chemistry at the TU Berlin free, Bohlmann was deemed the natural successor. He started work at the Institute for Organic Chemistry in 1959, and soon developed an impressive group of collaborators around him. The Society of German Chemists awarded him with the Otto Wallach prize in 1974. Despite having many international contacts and despite widespread recognition of his achievements, the prizes he received were limited to these German awards. He died on 23 September 1991.

Bohlmann's passion was for natural materials, their isolation, describing their structure and synthesising them. He was a leader in the skilful field of applying spectroscopic methods ("Bohlmann bands") in infra-red spectroscopy, as he was in the development of synthetic procedures and he was known to be an efficient organiser of his colleagues. Bohlmann was gifted with an extraordinary memory and had a great drive to achieve more, though the quality of his research was never compromised. Discipline and perseverance made him and his circle of colleagues into one of the world's most productive teams. They produced 1,453 publications.

in Braunschweig erkannte er die Compositen als besonders geeignete Untersuchungsobjekte. Die fast unerschöpfliche Vielfalt der Polyine und ihrer direkten Abkömmlinge begeisterte ihn dann sein ganzes Forscherleben lang. Isolierung, Strukturbestimmung und Synthese waren die eine Sache, das Erkennen der gegenseitigen Beziehungen, die Biogenese und Biotransformation die andere. Es erforderte, ganz der Bohlmannschen Denk- und Arbeitsweise entsprechend, die niedermolekularen Inhaltsstoffe in voller Breite zugänglich zu machen. Somit stellen Bohlmanns Arbeiten auch wichtige Beiträge zur chemischen Taxonomie dar. Bohlmann und C. Zdero begannen schon in den 60er Jahren mit der karteimäßigen Erfassung der Daten. In Verbindung mit dem Botanischen Garten Berlin entstand eine Arbeitsgruppe Biodiversitätsinformatik, die – längst als Datenbank betrieben – die „Bohlmann-Files" betreut. Ein zweites Interessengebiet betraf Verbindungsklassen wie die Chinolizidine, die ihn auf Grund ihrer Stereochemie und Konformationsanalyse fesselten.

Lit.: Ekkehard Winterfeld: Ferdinand Bohlmann (1921–1991) und sein wissenschaftliches Werk, in: Liebigs Annalen der Chemie 1994, S. I–X; Henry Laurent u. Rolf Bohlmann: Publikationen von Ferdinand Bohlmann 1948–1992, ibid. S. XI–XXXIV.

[M. E.]

Ferdinand Bohlmann (1921–1991)

Bohlmann's chief interests were terpenes and polyines and he found it particularly exciting when these substances occurred in nature. While still at Braunschweig he recognised the Compositae to be an especially suitable area to investigate. The almost inexhaustible variety of polyines and their direct derivatives went on to fascinate him right through his research career. The work consisted of isolation, determination of structure and synthesis on the one hand, while on the other it involved biogenetics, biotransformation and the recognition of reciprocal relationships. This demanded, in accordance with Bohlmann's thinking and working methods, making low-molecular material accessible in its full breadth. In this way, Bohlmann's work represents an important contribution to chemical taxonomy. In the 1960s Bohlmann and Christa Zdero started to compile a card index which became known as the Bohlmann files. In cooperation with Berlin's Botanical Gardens, a working group on biodiversity compiled a database of natural substances in the Compositae. Another field of interest for Bohlmann was quinolizidine alkaloids, which he was fascinated by due to their stereochemistry and conformational analysis.

Lit.: Ekkehard Winterfeld: Ferdinand Bohlmann (1921–1991) und sein wissenschaftliches Werk, in: Liebigs Annalen der Chemie 1994, p. I-X; Henry Laurent u. Rolf Bohlmann: Publikationen von Ferdinand Bohlmann 1948–1992, ibid. p. XI-XXXIV.

[M. E.]

Heinz Brauer wurde am 28.12.1923 in Oldenburg geboren. Nach Wehrdienst und Kriegsgefangenschaft studierte er 1949–1954 an der TU Hannover Maschinenbau. 1954–1959 am Max-Planck-Institut für Strömungsforschung in Göttingen tätig, wurde er 1956 an der TU Hannover zum Dr.-Ing. promoviert und habilitierte sich dort 1959 für Verfahrenstechnik. Ab 1960 war er Leiter der Wärmetechnik der Mannesmann AG in Duisburg.

1963 wurde Brauer auf den Lehrstuhl für Verfahrenstechnik an die TU Berlin berufen. Hier führte er die Mehrphasenströmung (also von Gemischen fester, flüssiger und gasförmiger Stoffe) und den Stoffaustausch mit und ohne chemische Reaktionen in die Lehre ein und bezog zunehmend auch die Prozesse der Wärmeübertragung und Biotechnologie mit ein. So schuf er ein umfassendes System der Transportprozesse einschließlich der chemischen und mikrobiellen Stoffumwandlungen. Schwerpunkte der Forschungen wurden die biologische Abwasserreinigung und die Luftreinhaltung.

Brauer führte schon in den 1960er Jahren computergestützte Übungen ein; in seinen Laborveranstaltungen „Innovationstechnik" wurden wissenschaftliche Grundlagenkenntnisse auf innovative Möglichkeiten untersucht. Auch nach der Emeritierung 1992 hält er noch Seminare an der TU ab, an der er mehrmals Dekan war und 73 Promotionen betreute. Brauers Verdienste um Verfahrenstechnik und Umweltschutz wurden mehrfach durch den Verein Deutscher Ingenieure, vom Bundesministerium für Forschung und Technologie und Privatstiftungen mit Preisen und Medaillen gewürdigt.

Seit den 1970er Jahren erwarben bei ihm bis zu 50% der Absolventen im „Europa-Studium" mit Anerkennung ausländischer Prüfungen ein Doppel-Diplom. Trotz politischer Schwierigkeiten für West-Berliner im Kontakt mit dem Ostblock engagierte sich Brauer schon in den 1970er

Heinz Brauer was born in Oldenburg on 28 December 1923. After undergoing service in World War II and imprisonment afterwards, he studied mechanical engineering at the Technische Universität of Hanover from 1949 to 1954. From 1954 to 1959 he worked at the Max Planck Institute for Flow Research in Göttingen. He took a doctorate in engineering at the Technische Universität of Hanover in 1956 and became a professor of process engineering there in 1959. In 1960 he became director of heat engineering at the Mannesmann company in Duisburg.

In 1963 Brauer took up the professorship of process engineering at the Technische Universität Berlin. In this post he introduced to the syllabus polyphase (i.e. mixed solid, fluid and gaseous materials) currents and the exchange of materials with and without chemical reactions and he increasingly included the processes of heat transfer and biotechnology. Thus he created an extensive system of transfer processes including chemical and microbial conversion. At the heart of the research he carried out were the biological treatment of waste water and air pollution control.

As early as the 1960s, Brauer had introduced computer-supported exercises and in his "Innovationstechnik" ("technological innovation") laboratory presentations he tested basic scientific principles for any innovative possibilities they may offer. He became professor emeritus in 1992 and continues to teach at the TU Berlin. He has held the post of dean several times and has supervised 73 students in their doctoral work. The German Association of Engineers, the Federal Ministry of Education and Research and private foundations have honoured Brauer's contributions to process engineering and environmental protection with various prizes and awards.

Since the 1970s up to 50% of the graduates who have worked with Brauer in the "Europa Studium" programme have achieved a double diploma

Jahren in der deutsch-polnischen Zusammenarbeit. Seit 1975 führte er mit Krakauer Kollegen Seminare über „Verfahrenstechnik und chemisches Apparatewesen" und „Umweltschutz in Ballungsgebieten" an der TU Berlin und der Politechnika Krakowska (TU Krakau) durch. Seit 1992 existiert ein gemeinsamer Studiengang „Umweltschutz", an dem Praktiker aus ganz Polen teilnehmen. Für all dies wurde er mit dem Ehrendoktor der Krakauer Universität 1988, dem Goldenen Ehrenzeichen 1995 und der Goldenen Ehrenmedaille 2003 der TU Krakau sowie dem polnischen Goldenen Orden für Umweltschutz 1998 ausgezeichnet. Seit 1985 pflegte Brauer auch intensive Kontakte und gemeinsame Seminare mit der Universität Miskolc/Ungarn, von der er 1991 ebenfalls einen Ehrendoktor erhielt.

Brauer schrieb mehrere Bücher, darunter fünf Lehrbücher, zahlreiche Buchbeiträge und über 200 Aufsätze in Fachzeitschriften, zudem gab er das Handbuch des Umweltschutzes und der Umwelttechnik Band V heraus. Seit 1971 hat er schwer zu beschreibende Vorgänge, etwa über Gasdispergierung und Feststoffbewegungen in mehrstufigen Wirbelschichten, in wissenschaftlichen Filmen für seine „Filmvorlesungen" veranschaulicht. Sein gesellschaftliches Engagement schlug sich in Mitgliedschaften von Vorständen und Beiräten nieder, er wirkte als Gastprofessor in mehreren Kontinenten und war nicht zuletzt Mitbegründer der Berliner Hochschulgesetze.

Lit.: Archiv der TU Berlin

[B. E.]

thanks to the recognition of foreign examinations. Despite political problems for West Berliners who had contact with Eastern Bloc countries during the 1970s, Brauer was engaged in German-Polish cooperation. In 1975 he started to hold seminars with Polish colleagues on process engineering and chemical apparatus studies and on environmental protection in metropolitan areas both at the TU Berlin and at the Politechnika in Krakow, Poland. A joint course on environmental protection has been offered since 1992, to which practitioners from all over Poland come.

For all of this work, he was given an honorary doctorate from the University of Krakow in 1988, the golden medal (1995) and the golden medal of honour (2003) from the Politechnika of Krakow, and in 1998 he was admitted to the Polish golden order for environmental protection. Since 1985, Brauer has also maintained close contact and held joint seminars with the University of Miskolc in Hungary, which awarded him an honorary doctorate in 1991.

Brauer has written several books, including five textbooks, and made numerous contributions to other publications, and he has published more than 200 articles for specialist journals. He also edited volume V of the handbook on environmental protection and environmental technology. Since 1971 he has used scientific films in his film-lectures in order to demonstrate topics that are difficult to describe, such as gas dispersion and the movement of solids in mutilevel fluid beds. His social commitments have been reflected in his membership of several boards of directors and advisory committees. Brauer has worked as a visiting professor on several continents and last but not least was a co-founder of the Berlin law on higher education.

Lit.: Archives of the TU Berlin

[B. E.]

Carl Dahlhaus (1928–1989)

Carl Dahlhaus wurde als Sohn eines Ingenieurs am 10.6.1928 in Hannover geboren. Schon früh musikbegeistert, erwirkte er für sich Klavierunterricht. Die höhere Schulbildung hingegen wurde zeitbedingt durch Dienste als Flakhelfer, im Arbeitsdienst und an der Front ersetzt; das Abitur konnte er erst nach dem Besuch eines Kriegsteilnehmerkurses ablegen. Durch einen Freund der Familie früh mit verbotener Literatur bekannt gemacht, begann er ungewöhnlich belesen 1947 zunächst das Studium der Rechte, dann der Musik- und Literaturwissenschaft in Göttingen und Freiburg. Noch als Student wurde er durch Empfehlung von Bertolt Brecht Dramaturg bei Heinz Hilpert (1890–1967) am Deutschen Theater in Göttingen, wo er an der Universität 1953 mit einer Arbeit über Josquin des Prés promoviert wurde.

1960 wechselte Dahlhaus als Musikredakteur an die Stuttgarter Zeitung. So konnte er zu Uraufführungen bedeutender Werke der Neuen Musik reisen, gleichzeitig forschte er über Renaissancemusik, Musikästhetik, -theorie und -analyse. 1962–1966 arbeitete er als Wissenschaftlicher Sachbearbeiter an der Universität Kiel, an der er sich 1966 mit „Untersuchungen über die Entstehung der harmonischen Tonalität" für Musikwissenschaft habilitierte. Nach kurzer Zeit an der Universität Saarbrücken wurde er als Nachfolger von Hans Heinz Stuckenschmidt (s.d.) 1967 auf den Lehrstuhl für Musikwissenschaft an die TU Berlin berufen, der er bis zu seinem Tod treu blieb; er lehnte alle Rufe anderer Universitäten ab, lehrte aber zwei Gastsemester in den USA (u.a. Princeton).

Dahlhaus verschaffte nicht nur als unermüdlicher Lehrer und geistvoller Diskutant seinem „Orchideenfach" eine ungewöhnliche Geltung. Er schrieb auch 25 Bücher, über 400 Aufsätze und weit über 150 Glossen, Musikkritiken und Buchrezensionen. Er gab das von ihm begründete Neue Handbuch der Musikwissenschaft, Pipers Enzyklopädie des Musiktheaters und die Ergänzungsbände zu Riemanns Musiklexikon heraus und war Mitherausgeber von vier Fachzeitschriften. Der Mitbegründer der Darm-

Carl Dahlhaus was born in Hanover the son of an engineer on 10 June 1928. He displayed an early interest in music and obtained piano lessons for himself. During World War II, his education was interrupted by his service as an anti-aircraft auxiliary, community work and at the front. He was able to complete his school exams only after attending a course for those engaged in combat. He became acquainted with banned literature at an early age through a family friend. By 1947, he was unusually well-read as he started studying law, and then music and literature in Göttingen and Freiburg. On the recommendation of Bertolt Brecht, he became a dramatic advisor at the Deutsches Theater in Göttingen while still a student. It was at the University of Göttingen that he took his doctorate in 1953 with a dissertation on the Renaissance composer Josquin des Prés.

In 1960 Dahlhaus became the musical editor of the Stuttgarter Zeitung newspaper. This enabled him to attend premieres of significant works in the New Music movement. At the same time, he researched Renaissance music, musical aesthetics, theory and analysis. Between 1962 and 1966 he worked as research associate at the University of Kiel, where he gained the grade of professor of musicology in 1966 with "Studies on the Origin of Harmonic Tonality." After a short period at the University of Saarbrücken he succeeded Hans Heinz Stuckenschmidt (q.v.) as professor of musicology at the TU Berlin, where he remained until his death. He turned down all job offers from other universities, though he did teach at Princeton as a visiting professor for two semesters.

As an inexhaustible teacher and a spirited debater Dahlhaus earned high esteem for his rather esoteric field of expertise. He wrote 25 books, more than 400 articles and over 150 glosses, pieces of criticism and book reviews. He co-founded and edited the Neue Handbuch der Musikwissenschaft, Pipers Enzyklopädie des Musiktheaters and supplementary volumes to Riemanns Musiklexikon. He also co-edited four academic journals.

städter Musiktage und zeitweilige Präsident der Gesellschaft für Musikforschung war Mitglied der Akademie für Sprache und Dichtung und des
Musikrats. Mit dem Großen Bundesverdienstkreuz mit Stern und der Auf-
nahme in den Orden Pour le mérite wurde der Schriftsteller, Wissenschaftler und Organisator gewürdigt.

Dahlhaus schuf der Musikwissenschaft eine neue Grundlegung,
indem er sie methodisch bereicherte, Geschichtstheorie, Musikästhetik,
Musiktheorie und –analyse miteinander verbindend. Seine Studien zur
Vorgeschichte der Neuen Musik – er hatte Schönbergs Werke mit herausgegeben – führten Dahlhaus zur „Musik des 19. Jahrhunderts" (1980),
die er der Musikwelt nicht nur mit der Richard-Wagner-Gesamtausgabe,
sondern auch musiktheoretisch und -ästhetisch neu erschloss. Zunehmend historisch denkend, fasste er Musik, bei voller Anerkennung ihrer
Autonomie, als Kunstform auf, die an den das jeweilige Zeitalter bestim-
menden Ideen teilhat.

Umfassend gebildet, witzig, rational und mit entschiedenem Urteil
bereicherte er jede Diskussion. Seine Liebenswürdigkeit, ja Lebensfreu-
de ließen viele nichts von seinen schweren Erkrankungen ahnen, denen
er seine fundamentalen, immer geistreichen Werke abgerungen hatte. Er
erlag am 13.3.1989 in Berlin seinem langen Leiden.

Lit.: Hrsg. H. Danuser u.a.: Das musikalische Kunstwerk. Geschichte, Ästhetik, Theorie.
Festschrift Carl Dahlhaus zum 60. Geburtstag, Laaber, 1988. – Rudolf Stephan: Carl
Dahlhaus (1928–1989), in: Die Musikforschung 42 (1989), S. 203–206.

[B. E.]

He was a co-founder of the Darmstädter Musiktage music festival and served intermittently as president of the Gesellschaft für Musikforschung (Society for Music Research) and was a member of the Akademie für Sprache und Dichtung (Literary Academy) and of the Musikrat (Music Council). As a writer, academic and administrator, the German state awarded him the Order of Merit of the Federal Republic and he was also a member of the Orden Pour le Mérite.

Dahlhaus set out a new basis for musicology by enriching its methodology to include historical theory, musical aesthetics, musical theory and musical analysis. His study of the prehistory of the New Music – he took part in the publication of the works of Arnold Schönberg – led Dahlhaus to his 1980 work "Musik des 19. Jahrhunderts" ("Nineteenth-Century Music"), a musical period which he opened up to the music world not only with the complete works of Richard Wagner, but also in terms of musical theory and musical aesthetics. His thought grew increasingly historical, and, while he recognised the autonomy of music, he also conceived of it as an artform that participates in the ideas of the era in which it takes place. Widely educated, humourous, clear-thinking and endowed with a fine sense of judgement, Dahlhaus was a stimulating presence in any discussion

His amiable nature and joie de vivre left many unaware of his serious illnesses, on which he drew for his fundamental, spiritually-rich works. He died in Berlin on 13 March 1989 after a long illness.

Lit.: H. Danuser (ed.) et al.: Das musikalische Kunstwerk. Geschichte, Ästhetik, Theorie. Festschrift Carl Dahlhaus zum 60. Geburtstag, Laaber, 1988. – Rudolf Stephan: Carl Dahlhaus (1928–1989), in: Die Musikforschung 42 (1989), p. 203–206

[B. E.]

Hanswerner Dellweg (*1922)

Der am 19.2.1922 in Aschaffenburg geborene *Hanswerner Dellweg* studierte ab 1940 Chemie an den Universitäten Köln und Frankfurt/M. 1941 wurde er als Soldat eingezogen und konnte das Studium erst nach französischer Kriegsgefangenschaft 1947 in Heidelberg fortsetzen. Dort wurde er 1952 mit einer Arbeit über Nucleinsäuren promoviert: Ein Jahr vor Aufdeckung der Doppelhelix-Struktur von DNS isolierte er die Desoxinucleoside aus roher DNS, stellte sie präparativ dar und untersuchte ihr Verhalten im Stoffwechsel von Bakterien.

Seit 1948, als Werkstudent bei den Aschaffenburger Zellstoffwerken, hatte Dellweg sich mit der Vergärung und Verhefung von Sulfitablaugen sowie der Technik zur Gewinnung von Sulfitsprit befasst und dabei neue Analysemethoden entwickelt. Nach der Promotion trat er nun als Wissenschaftlicher Mitarbeiter in das Biochemische Labor dieser Firma ein. Hier erarbeitete er Methoden der Anreicherung und Reindarstellung von Corrinoiden aus Faulschlamm, einer Substanzgruppe, zu der das Vitamin B_{12} gehört. Ab 1956 leitete Dellweg eine halbtechnische Versuchsanlage zur Gewinnung von Vitamin B_{12}; dabei gelang ihm mit Hilfe von Colibakterien die in-vitro-Synthese von Analogen des Vitamins B_{12}, dessen Konstitution Todd u.a. erst 1955 aufgeklärt hatten (Totalsynthese 1971 durch Eschenmoser).

1960 ging Dellweg als Wissenschaftlicher Rat an das Institut für Therapeutische Biochemie der Universität Frankfurt. Hier forschte er u.a. über strahlenchemische Veränderungen von Nucleinsäuren und habilitierte sich 1965 für Therapeutische Biochemie.

1967 wurde Dellweg auf den Lehrstuhl für Gärungstechnologie – später Biotechnologie – an die TU Berlin berufen, der außerhalb der TU im Institut für Gärungsgewerbe (Seestraße) untergebracht war. Seit 1968 wissenschaftlicher Direktor des Instituts für Gärungswissenschaf-

Hanswerner Dellweg was born on 19 February 1922 in Aschaffenburg. He started to study chemistry in 1940 at the universities of Cologne and Frankfurt. He was conscripted by the army in 1941 and was able to complete his studies in Heidelberg only in 1947 after being held as a prisoner of war by the French. At Heidelberg he earned a doctorate in 1952 with a dissertation on nucleic acids: one year before the discovery of the double-helix structure of DNA, he isolated deoxynucleoside from untreated DNA, made a preparation of it and examined its behaviour in the metabolism of bacteria.

Dellweg started work experience at a pulp factory in Aschaffenburg in 1948. There he learned about fermentation and the behaviour of yeast in sulphite lyes and about the technology used to extract sulphite ethanol and he developed new methods of analysis along the way. After earning his doctorate he took a post as a scientist in the biochemical laboratory of the same company. There he worked out methods for the enriching and isolation of corrinoids out of waste effluent; corrinoids are a group of substances that include Vitamin B_{12}. From 1956 Dellweg directed a pilot plant dedicated to the extraction of Vitamin B_{12}. His success owed something to the help of coli bacteria, which are the in-vitro synthetics of analogues of Vitamin B_{12}, whose constitution had been elucidated by Alexander Todd et al. as recently as 1955 (the total synthesis of Vitamin B_{12} was achieved by Albert Eschenmoser in 1971).

In 1960 Dellweg became a scientific advisor in the Institut für Therapeutische Biochemie (Institute for Therapeutic Biochemistry) of the University of Frankfurt. Among the subjects he researched there was radiochemical variations of nucleic acids. He qualified as a lecturer in Therapeutic Biochemistry in 1965.

In 1967 Dellweg became professor of Gärungstechnologie, later Biotechnologie (fermentation technology, later biotechnology) at the

ten, untersuchte er klassische Probleme der Gärungs-Chemie bis hin zur vollständigen Kohlehydratzusammensetzung in fermentierenden Würzen, Stoffwechsel in Hefen und Enzymkinetik. Gegen großen Widerstand gelang es Dellweg in den 1970er Jahren, den ersten eigenen Studiengang Biotechnologie Deutschlands an der TU zu etablieren. Als Wissenschaftlicher Leiter des Instituts für Biotechnologie sah er seine Aufgabe darin, Ingenieuren und Verfahrenstechnikern biologisches Denken und Grundkenntnisse der Mikrobiologie zu vermitteln, so dass sie biotechnologische Aufgaben ohne Doppelstudium lösen können. Er las über Biochemie, Industrielle Mikrobiologie, Regulation des mikrobiellen Stoffwechsels und Reaktionskinetik der Fermentation. Er forschte über Hefen zur Gewinnung von Einzellerprotein aus Erdöl-Fraktionen, Methanol als Kohlenstoffquelle für Hefen und Bakterien sowie Grundlagen der anaeroben Abwasserreinigung. Seit 1975 versah er den Lehrauftrag „Verfahren der Biotechnologie" an der Freien Universität Berlin.

Dellweg organisierte in Berlin die Symposien „Technische Mikrobiologie" der Jahre 1970, 1975, 1979 und 1982, wurde 1976 Vice-Chairman der Fermentation Commission und 1979 Chairman der Commission of Fermentation Industries der IUPAC und gab den Ergänzungsband „Biochemie" des Römpp-Chemielexikons heraus. 1977 wurde er zum auswärtigen Mitglied der Finnischen Akademie der Technischen Wissenschaften gewählt. Seit der Emeritierung 1990 lebt Dellweg im Fichtelgebirge.

[B. E.]

TU Berlin. The department was located outside the main campus, in the Institut für Gärungsgewerbe (Institute for Commercial Fermentation) in Seestraße. He became scientific director of the Institut für Gärungswissenschaften (Institute for Fermentation Sciences) in 1968, and studied classic problems of fermentation chemistry, such as the composition of carbohydrates in fermentation, metabolism in yeast and enzyme kinetics. In the 1970s, Dellweg prevailed against much opposition to establish the first full course on biotechnology in Germany. As the scientific director of the Institut für Biotechnologie he saw it as his job to educate engineers and process technicians in the biological principles of microbiology in order that they would be able to take on biotechnological tasks without having to take two degrees. He lectured on biochemistry, industrial microbiology, the regulation of microbial metabolism and the reaction kinetics of fermentation. He researched yeast to extract unicellular proteins from certain types of petroleum, methanol as a source of carbon for yeast and bacteria, as well as the basic principles of the anaerobic treatment of waste water. From 1975 he gave a course on processes in biotechnology at the Free University of Berlin.

Dellweg organised the Technische Mikrobiologie (Technical Microbiology) symposiums in Berlin in 1970, 1975, 1979 und 1982. In 1976 he became vice-chairman of the Fermentation Commission and in 1979 Chairman of the Commission of Fermentation Industries of the International Union of Pure and Applied Chemistry. He also edited the supplementary volume on biochemistry for the Römpp-Chemielexikon chemical dictionary. In 1977 he was made a foreign member of the Finnish Academy of Technical Sciences. He has lived in the Fichtelgebirge region of Bavaria since becoming professor emeritus in 1990.

[B. E.]

Ernst Fiala (*1928)

Josef Herbert Ernst Fiala wurde am 2.9.1928 in Wien geboren. Er studierte an der TH Wien Maschinenbau. Nach dem Diplom war er dort von 1952 bis 1954 Assistent am Institut für Verbrennungskraftmaschinen und Kraftfahrwesen. 1954 wurde er mit einer Arbeit über Seitenführungskräfte am rollenden Luftreifen zum Dr. techn. promoviert. Seine Berufslaufbahn begann Fiala als Versuchs- und Entwicklungsingenieur bei der Daimler-Benz AG in Sindelfingen, zuletzt war er dort Leiter der Entwicklungsabteilung für PKW-Aufbauten. 1963 wurde er als Ordinarius an das Institut für Kraftfahrzeuge der TU berufen, wo er bis 1970 auch mit der Leitung des Technischen Prüfzentrums für Kraftfahrzeuge betraut war. Als Hochschullehrer zeigte er die seltene Fähigkeit, komplizierte technische Vorgänge in leicht verständlicher Weise erläutern zu können. Das führte auch dazu, daß man ihn nebenamtlich in das Direktorium der Hamburger Verkehrsakademie berief.

Studentenunruhen, Hochschulreform und das neue Universitätsgesetz boten einen nicht zu unterschätzenden Anstoß zum Weggang aus Berlin. Am 1.10.1970 übernahm Fiala den Zentralbereich Forschung und im März 1972 kommissarisch die Leitung des Vorstandsbereichs Forschung und Entwicklung der Volkswagen AG. Seit dem 1.2.1973 wurde er zum Vorstandsmitglied bestellt. In seine Zeit fiel die Einführung des VW-Golf, der 1974 auf den Markt kam, den legendären „Käfer" ablöste und zu einem der erfolgreichsten Modelle wurde. Neben der Weiterentwicklung des Golf, der weiteren PKW-Programme und des VW-Nutzfahrzeugprogramms musste er sich mit den Folgen der Ölkrise, dem wachsenden Umweltbewusstsein sowie den gestiegenen Sicherheitsanforderungen auseinandersetzen. Er hatte Anteil an der Entwicklung des Dreiwegekatalysators, der Senkung der C_w-Werte, des Dieselmotors für Mittelklassewagen und der spurstabilisierenden Lenkung.

Nach Vollendung des 60. Lebensjahres gab Fiala seinen Posten auf, um sich intensiver der wissenschaftlichen und publizistischen Arbeit zu

Josef Herbert Ernst Fiala was born in Vienna on 2 September 1928. He studied mechanical engineering at the TH Wien. After taking a master's degree, he worked from 1952 to 1954 as an assistant in the Institute for Combustion Engines and Automotive Engineering. He was awarded a doctorate in 1954 for his work on lateral force in rolling pneumatic tyres. Fiala started working as a research and development engineer at Daimler-Benz in Sindelfingen, where he eventually ended up as director of the research department for car bodies. In 1963 he became professor at the TU Berlin's Institute for Automotive Engineering, where he remained until 1970, by which time he had become the director of the technical testing centre. As a college teacher, he displayed the rare gift of being able to explain complicated technical details in an easily understood way. This also led to his being invited to join the directorate of the Hamburger Verkehrsakademie (Hamburg School of Shipping and Transportation).

A combination of student unrest, reform of third-level education and the new university law provided Fiala with persuasive reasons to leave Berlin. On 1 October 1970 he took charge of the Central Research Unit at Volkswagen and in March 1972 he took over the directorship of the research and development board at the company.

He joined the Volkswagen board of directors in February 1973. During his tenure, the VW Golf was brought to the market (1974), replacing the legendary Beetle and becoming one of the world's most sucessful models. Alongside the development of the Golf, other passenger car projects and VW's utility vehicles programme, Fiala also had to deal with the effects of the oil crisis, growing environmental awareness and rising safety standards. He played a part in developing the three-way catalytic converter, lowering drag coefficient, developing diesel motors for mid-range cars and developing steering with tracking stability.

widmen. Als Berater blieb er VW verbunden, und als Honorarprofessor an der TU Wien lehrte er das Fach „Wechselbeziehungen zwischen Mensch und Fahrzeug". Zudem ist er Mitglied in mehreren Aufsichtsräten.

In dem Buch „Wieviel Auto braucht der Mensch" (1990) setzt er sich mit dem Möglichen und Machbaren auseinander. Er vertritt eine Gegenposition zur Technik- und Wachstumsskepsis, setzt der Studie „Grenzen des Wachstums" des Club of Rome ein „Wachstum ohne Grenzen. Wohlstand durch globales Wachstum" (mit Erich Becker-Boost, 2001) entgegen, worin die These vertreten wird, dass nicht der internationale Wettbewerb sondern der Mangel an Wettbewerb die Ursache von Armut sei. Zur Lösung der Weltprobleme sei hingegen mehr, nicht weniger Technik vonnöten. Er veröffentlichte zahlreiche Fachaufsätze, und es wurden ihm mehr als 100 Patente erteilt.

Fiala erhielt zahlreiche Ehrungen, darunter die Ehrendoktorwürde der Universitäten Heidelberg und Kragujevac (Serbien). Seine Verdienste um Österreich wurden mit der Verleihung des Goldenen Ehrenzeichens gewürdigt.

Lit: Munzinger Archiv; TU-Archiv

[M. E.]

At the age of 60, Fiala resigned from his full-time post so that he could devote himself more intensively to scientific work and writing. He stayed on as a consultant for VW and was made honorary professor at the University of Vienna, where he taught a course on the interrelationship between humans and vehicles. He is also a member of several boards of directors.

In his 1990 book „Wieviel Auto braucht der Mensch", Fiala addresses what is possible and what can be achieved. He takes a position against those who are sceptical of the benefits of technology and growth. In 2001 he countered the arguments set out in the Club of Rome study "The Limits of Growth" with "Wachstum ohne Grenzen. Wohlstand durch globales Wachstum" (Limitless Growth: Prosperity through Global Growth), which he co-wrote with Erich Becker-Boost. The thesis proposed in this book is that it is not international competition but the lack of international competition that is the root cause of poverty. The book argues that more and not less technology is needed to find a solution to the world's problems. Fiala has published many specialist articles and has been awarded more than 100 patents.

Fiala has received many awards, including honorary doctorates from the universities of Heidelberg and Kragujevac, Serbia. His native Austria has conferred him with the Grand Decoration of Honour.

____

Lit.: Munzinger Archive; TU Archive

[M. E.]

Hermann Föttinger wurde am 9. Februar 1877 in Nürnberg geboren. Nach bestandenem Abitur schrieb er sich 1895 für das Studium der Elektrotechnik an der Technischen Hochschule München ein, wandte sich aber noch während seines Studiums auch maschinenbaulichen Forschungsproblemen zu. Nach Abschluss des Studiums 1899 fand er seine erste Anstellung bei der Stettiner Schiffsbau-Anstalt Vulcan, wo er in der Versuchsanstalt für Schiffsantriebe arbeitete. Hier entwickelte er zunächst ein Torsionsdynamometer, das direkt auf die Schiffswelle montiert werden konnte und bei laufender Maschine Drehmoment, Leistung und Drehschwingung maß. Für das Torsionsdynamometer erhielt Föttinger das erste von über einhundert Patenten seines Lebens, und es war Gegenstand seiner Dissertation, mit der er 1904 in München promovierte.

Zur gleichen Zeit beschäftigte sich Föttinger mit dem Problem der Drehmomentwandlung. Man wollte auch Dampfturbinen zum Antrieb von Schiffen einsetzen. Anders als Dampfmaschinen liefern Dampfturbinen aber erst bei hohen Umdrehungszahlen optimale Leistung. Man brauchte deshalb ein Getriebe, das die hohen Drehzahlen auf die für den Propeller erforderliche niedrige Drehzahl herabsetzte, mechanische Zahnradgetriebe konnte man jedoch für die hohen Leistungen von Schiffsantrieben noch nicht bauen. Auch Föttingers Idee eines hydrodynamischen Wandlers über die Kopplung einer Pumpe und einer Turbine schien zunächst aussichtslos, da die Strömungsverluste in den Leitungen zu einem sehr niedrigen Wirkungsgrad des Getriebes geführt hätten. Erst die Idee, Pumpe und Turbine in einem Gehäuse zusammen zu legen, versprach einen akzeptablen Wirkungsgrad. Eine Versuchsausführung lieferte den erstaunlich günstigen Spitzenwirkungsgrad von 83 %.

Föttinger war inzwischen bei Vulcan zum Chef des Versuchswesens aufgestiegen. 1909 wurde er auf das Ordinariat für Schiffsmaschinenbau der TH Danzig berufen. Föttinger widmete sich hier insbesondere der

Hermann Föttinger was born in Nuremberg on 9 February 1877. Having finished his school exams, he enrolled in a course in electrical engineering at the TH Munich. While studying there, he devoted much time to solving problems in mechanical engineering. Föttinger graduated in 1899 and found his first job at the Vulcan shipbuilding company in Stettin (Szczecin). It was while working there that he first developed his torsion dynanometer, which could be mounted directly onto the marine shaft of a ship and was used to measure the torque, output, and torsional vibration of the running engine. The torsion dynanometer earned Föttinger his first patent (he was to take out the more than one hundred during his career) and it was the subject of his doctoral dissertation, which he completed in Munich in 1904.

At the same time, Föttinger grew interested in the problem of torque conversion. In the early years of the twentieth century, the steam turbine was being considered as a possible means of marine propulsion. Unlike steam engines, steam turbines perform at their best only at a high rate of revolutions. What was needed was a drive shaft that would reduce this high spin rate to the low rate necessary for powering a propeller. Gear mechanisms capable of delivering the high output required for ship engines had not yet been built. Föttinger's idea of developing a hydro-dynamic converter across the coupling joining the pump and the turbine held out little hope at first due to the very high rates of flow loss in the pipes, which meant that the engine worked only at very low efficiency. It seemed to Föttinger that an acceptable level of efficiency could be achieved only by combining the pump and turbine in one housing. This was tested, and gave an extremely impressive efficiency rate of 83%.

In the meantime, Föttinger was in charge of research and development at Vulcan. In 1909 he was awarded the chair of Marine Engineering at the TH in Danzig (Gdansk). In this post Föttinger devoted himself in particular to fluid mechanics, a field of study in which there was a wide gap between

Strömungslehre, auf diesem Gebiet klaffte zwischen theoretisch-klassischer Hydrodynamik und empirisch-angewandter Hydraulik noch eine riesige Lücke. Schon bei Vulcan hatte er mit Untersuchungen zur Kavitation begonnen, die erhebliche Zerstörungen an Schiffspropellern verursachen kann. In Danzig wies er nun nach, dass diese Zerstörungen durch die mechanischen Kräfte bei der Implosion der Dampfblasen verursacht werden, die durch die Kavitation entstehen. Grundlegende Forschungsbeiträge lieferte Föttinger auch auf den Gebieten der Propellertheorie, der Wechselwirkung zwischen Schiff und Propeller und der Simulation der Strömung um Schiffskörper.

1924 wurde Föttinger an die TH Berlin auf den Lehrstuhl für Allgemeine Strömungslehre und Turbomaschinen berufen, den ersten Lehrstuhl in Deutschland, der die Bezeichnung Strömungslehre trug. Er machte sich engagiert an den Aufbau eines neuen Instituts und sammelte eine Schar begabter Mitarbeiter um sich, die er zu Forschungen auf verschiedenen Gebieten der Strömungslehre anregte. In Berlin entstanden Forschungsbeiträge zur Spülströmung in Motoren, der Rauchgasführung in Kesseln sowie über die Kohlenstaubturbine. Föttinger starb 68-jährig in den letzten Kriegstagen am 28. April 1945 in Berlin, nachdem er in der Nähe seines Wohnhauses von einem Bombensplitter getroffen worden war.

Lit.: Föttinger-Getriebe. Vorträge auf dem Festkolloquium anlässlich der 100. Wiederkehr des Geburtstages von Hermann Föttinger, Fortschritt- Berichte der VDI-Zeitschriften, Reihe 7 Nr. 45, 1977

[J. Z.]

classic theoretical hydrodynamics and empirical, applied hydraulics. At Vulcan, he had already started looking into the problem of cavitation, which can cause serious damage to ship propellers. At Danzig, he demonstrated that this damage was caused by the mechanical power produced by the implosion of air bubbles, which are created by cavitation. Föttinger also conducted research that provided the basis for fields of activity such as propeller theory, the interaction between ship and propeller and the simulation of the movement of fluid around the hull of a ship.

In 1924, Föttinger became professor of General Fluid Dynamics and Turbomechanics at the TH Berlin. It was the first chair of fluid dynamics in Germany. He devoted his energy to creating a new institute and gathered around him a group of gifted colleagues, whom he inspired to investigate various aspects of the field of fluid dynamics. The institute conducted research in scavenging flow in motors, the extraction of flue gas from boilers, and the coal-dust turbine. In the final days of World War II, Föttinger was struck by bomb shrapnel near his home in Berlin. He died on 28 April 1945, at the age of 68.

Lit.: Föttinger-Getriebe. Vorträge auf dem Festkolloquium anlässlich der 100. Wiederkehr des Geburtstages von Hermann Föttinger, Fortschritt- Berichte der VDI-Zeitschriften, Reihe 7 Nr. 45, 1977.

[J. Z.]

Johannes (Hans) Wilhelm Geiger wurde am 30. September 1882 in Neustadt an der Weinstraße geboren. Sein Vater Wilhelm Ludwig Geiger (1856–1943) war ein angesehener Iranistiker und Indologe. 1901 immatrikulierte sich Geiger an der Universität Erlangen für Mathematik und Physik, leistete aber in den ersten zwei Semestern noch seinen Militärdienst ab. Das Sommersemester 1904 verbrachte er an der Universität München, anschließend kehrte er nach Erlangen zurück. Im Herbst 1904 legte er die erste Lehramtsprüfung für Mathematik und Physik ab, im Frühjahr 1906 folgte die zweite. Im Sommer 1906 promovierte er mit einer Arbeit über „Strahlungs-, Temperatur- und Potentialmessungen in Entladungsröhren mit starken Strömen".

Geiger erhielt eine Assistentenstelle bei Arthur Schuster (1851–1934) am physikalischen Institut der Universität Manchester. In den Semesterferien des Sommers 1907 kam Ernest Rutherford (1871–1937) als Nachfolger von Schuster nach Manchester und bot Geiger an, weiter als Assistent in Manchester zu bleiben. Gemeinsam mit Rutherford veröffentlichte Geiger 1908 eine Arbeit über eine elektrische Zählmethode für Alphateilchen. Aus der genauen Zahl der Alphateilchen, die von einer gegebenen Menge Radium abgegeben werden, kann man direkt die Loschmidtsche Zahl ableiten, die damals nur sehr ungenau und aus indirekten Methoden ermittelt werden konnte. Im gleichen Jahr folgte noch, diesmal von Geiger allein, eine Arbeit mit dem Nachweis der statistischen Natur des radioaktiven Zerfalls. Gemeinsam mit Ernest Marsden (1889–1970) beobachtete er 1909, dass Alphateilchen an Metallfolien um Winkel von 90° und mehr gestreut werden können.

Ende 1912 verließ Geiger Manchester, um eine Stelle an der Physikalisch-Technischen Reichsanstalt (PTR) anzutreten. Hier sollte Geiger ein Laboratorium für Radioaktivität aufbauen und die nötigen Eich- und Prüfvorrichtungen erstellen. Gemeinsam mit James Chadwick (1891–1974), den Geiger aus Manchester mit an die PTR gebracht hatte, entwi-

Johannes (Hans) Wilhelm Geiger was born in Neustadt
an der Weinstraße on 30 September 1882. His father
Wilhelm Ludwig Geiger (1856–1943) was a highly respected scholar of
Persian and Indian culture. Hans Geiger enrolled at the University of Erlangen in 1901, where he studied mathematics and physics, though he did
his compulsory military service in the first two semesters. He spent the
summer semester of 1904 at Munich University, after which he returned
to Erlangen. He took his first lectureship exams in autumn 1904 and the
next exam followed in early 1906. He took his doctorate in the summer
of 1906 with a paper on measurements of radiation, temperature and
potential in high-flow discharge tubes.

Geiger took up a post as assistant to Arthur Schuster (1851–1934)
at the Physics Institute of the University of Manchester. In the summer
of 1907, Schuster was succeeded by Ernest Rutherford (1871–1937), who
asked Geiger to stay on as his assistant. Geiger and Rutherford published
a paper on methods for counting alpha particles in 1908. They found that
by working out the precise number of alpha particles that can be derived
from a given amount of radium, one can directly deduce the Avogadro
number, which at the time could be determined only very imprecisely and
using indirect methods. Also in 1908, Geiger was the sole author of a
paper giving proof of the statistical nature of radioactive decay. Together
with Ernest Marsden (1889–1970), in 1909 Geiger observed the reflection
of alpha particles from various metal foils.

At the end of 1912 Geiger left Manchester to take up a post at
the Physikalisch-Technische Reichsanstalt (German National Institute for
Science and Technology) in Berlin. There he built a lab for the study of
radioactivity and for calibration and control devices. With James Chadwick (1891–1974), whom Geiger had brought with him from Manchester,
he developed a needle counter. World War I interrupted Geiger's activities at the Physikalisch-Technische Reichsanstalt until 1919. In 1924,

ckelte er 1913 einen Spitzenzähler. Der Erste Weltkrieg unterbrach Geigers Tätigkeit an der PTR, erst 1919 konnte er sie fortsetzen. Bereits vor dem Krieg war Walther Bothe (1891–1957) zu Geiger an die PTR gekommen. 1924 begannen sie mit Versuchen zur Untersuchung des Compton-Effekts und entwickelten die Koinzidenzmethode.

1924 habilitierte sich Geiger an der Berliner Universität mit einer Sammlung von Arbeiten über die Reichweitemessung von Alphateilchen. 1925 folgte er einem Ruf an die Universität Kiel. Hier entwickelte er zusammen mit seinem ersten Doktoranden Walter Müller (1905–1979) das Proportionalzählrohr, mit dem man nicht nur radioaktive Teilchen und Gamma-Strahlung messen konnte, sondern auch deren Energie bestimmen konnte. Durch das Geiger-Müller-Zählrohr ist Geiger auch außerhalb der Physik zu einem berühmten Mann geworden.

1929 ging Geiger nach Tübingen und schließlich 1936 wieder nach Berlin, wo er zum Direktor des physikalischen Instituts der Technischen Hochschule ernannt wurde. In Berlin hat er die energiereiche Höhenstrahlung untersucht. Im November 1936 übernahm Geiger die Redaktion der Zeitschrift für Physik und wurde im gleichen Monat zum Mitglied der Berliner Akademie gewählt. Geiger starb am 29. September 1945 in Potsdam.

Lit.: Edgar Swinne: Hans Geiger. Spuren aus einem Leben für die Physik, Berlin: ERS-Verlag 1991

[J. Z.]

Geiger and Walther Bothe (1891–1957), who had joined Geiger at the Physikalisch-Technische Reichsanstalt before the war, began experiments investigating the Compton effect and they developed the technique of coincidence counting.

In 1924 Geiger became professor at Berlin University on the strength of his body of work on the measurement of alpha particles. In 1925, he moved to the University of Kiel, where he and his first doctoral student Walter Müller (1905–1979) developed the proportional counter tube, which measures not only radioactive particles and gamma rays, but also their energy. The Geiger counter, as the device became known, made Geiger a household name.

Hans Geiger moved to Tübingen in 1929 and finally to Berlin in 1936, where he was made director of the Physics Institute of the TH. In Berlin, he researched high energy cosmic radiation. In November 1936 he became editor of the Zeitschrift für Physik and in the same month was made a member of the Berlin Academy. Geiger died in Potsdam on 29 September 1945.

Lit.: Edgar Swinne: Hans Geiger. Spuren aus einem Leben für die Physik, Berlin: ERS-Verlag 1991

[J. Z.]

Felix Genzmer (1856–1929)

Als Sohn eines Juristen wurde *Felix Genzmer* am 22.11.1856 in Labes (Pommern) geboren. Er studierte an den THs Hannover und Stuttgart und begann 1880 seine praktischen Lehrjahre bei der Reichseisenbahn von Elsaß-Lothringen. Hier wurde er maßgeblich durch den Architekten Eduard Jacobsthal (1839–1902) beeinflusst. 1887 schied er dort aus und konnte in den folgenden drei Jahren als Assistent des städtischen Hochbauamtes in Köln einige kleinere Entwürfe ausführen und wurde danach als Stadtbaumeister in Hagen (Westfalen) vor allem mit der Provinzial-Gewerbeschule auch überregional bekannt. Seit 1894 Stadtbaumeister von Wiesbaden konnte er in dem nun größeren Wirkungskreis alle seine Begabungen zeigen. Neben einigen öffentlichen Bauten erregte vor allem der Foyeranbau des Wiesbadener Hoftheaters (1902), veranlasst durch die alljährlichen Kaiserfestspiele, Aufsehen. In Wiesbaden boten sich zahlreiche Möglichkeiten, sein dekorativ-ornamentales Können zur Wirkung zu bringen. Genzmers wesentliche Leistung bestand jedoch darin, die rege Bau- und Umbautätigkeit in der prosperierenden Kurstadt in ein umfassendes städtebauliches Konzept einzufügen. Seine Verdienste wurden 1901 mit dem Titel Kgl. Baurat gewürdigt.

1903 folgte er dem Ruf an die TH Berlin auf die Professur für Städtebau und farbige Dekoration an der Abt. Architektur, die in dieser Form auf seinen Wunsch hin eingerichtet wurde. Sie war die erste ihrer Art in Preußen. Im Nebenamt war er von 1903 bis 1920 Architekt der Kgl. bzw. Staatstheater in Berlin, beteiligte sich an zahlreichen Wettbewerben und veröffentlichte zahlreiche wissenschaftliche Arbeiten, die auch in der Gegenwart nicht ohne Aktualität sind. 1905 wurde er zum Geh. Hofbaurat ernannt.

Gemeinsam mit seinem Kollegen Josef Brix (1859–1943), der das Fach Städtebau in der Abteilung Bauingenieurwesen vertrat, gründete Genzmer in WS 1907/08 das Seminar für Städtebau, Siedlungs- und

Felix Genzmer was born in Labes (Pomerania) on 22 November 1856, the son of a lawyer. He studied at the Technische Hochschule in Hanover and at its equivalent in Stuttgart, before beginning his practical training on the Alsace Lorraine state railways in 1880. Here he was considerably influenced by the architect Eduard Jacobsthal (1839–1902). He left in 1887 and in the following three years was able to develop some smaller designs as an assistant at the Municipal Building Surveyor's Office in Cologne. Later he was to achieve renown beyond the region through his work as architect in Hagen (Westphalia), particularly with his design for the province's vocational school building. He was able to display his architectural talents to a larger public after becoming the municipal architect in Wiesbaden (1894). Alongside a number of public buildings, it was above all his entrance hall to the Wiesbaden court theatre (1902), which was built to commemorate the annual theatre festival held in honour of the Kaiser that excited interest. In Wiesbaden, he was given many opportunities to effectively demonstrate his decorative and ornamental abilities. However, Genzmer's main achievement consisted of integrating the active construction and reconstruction activity going on in the flourishing spa town city with a comprehensive town planning concept. His services were honoured in 1901 by the award of the title of Königlicher Baurat (Town Planner by Royal Appointment).

In 1903 he followed the call to the TH Berlin as "Professor for Town Planning and Use of Colour in Decoration" at the Architectural Department, which had been established in this form at his request. It was the first of its kind in Prussia. Simultaneously, from 1903 to 1920 he also worked as an architect for the Royal and National Theatres in Berlin, took part in numerous competitions and published many scientific works, which are still of relevance to this day. In 1905 he was appointed senior town planner to the royal court.

Wohnungswesen. Beide beteiligten sich 1910 erfolgreich am Wettbewerb Groß-Berlin. Weder dieser noch andere damals ausgezeichnete Pläne zu diesem Großprojekt waren realisierbar. Zuvor (1908) errangen beide den ersten wie auch den dritten Platz im Wettbewerb Gartenstadt Frohnau bei Berlin, die dann im wesentlichen nach ihren Vorschlägen errichtet wurde und noch immer zu den sehenswerten Teilen Berlins zählt. Zahlreiche weitere Bebauungspläne zeigen, daß es Genzmer bei allem immer auch um die Einzelheiten ging. So wies er der Behandlung städtebaulicher Einzelaufgaben und der Architektur im Städtebau erhebliche pädagogische Bedeutung zu. Von den kleineren Arbeiten Genzmers sind die Büffelhäuser im Zoologischen Garten erwähnenswert.

Als beste und reifste Leistung im Privathausbau gilt sein Landhaus in Dahlem (Podbielskiallee), wo er auch sein Talent im Entwurf von Möbeln, Beleuchtungskörpern und Verglasungen unter Beweis stellte. In seinen späteren Jahren übte er zahlreiche Ämter als Preisrichter bei Wettbewerben und als Obergutachter aus. Die TH verlieh ihm 1925 die Ehrenmitgliedschaft. Genzmer verstarb am 6.8.1929 in Dahlem.

Lit.: F. Schultze: Zum 70. Geburtstag von Felix Genzmer, in: Zentralblatt der Bauverwaltung 46 (1926), S. 513; Erich Blunck: Felix Genzmer; ein Bild seines Wirkens zu seinem 70. Geburtstag, in: Deutsche Bauzeitung 60 (1926), S. 753–760.

[M. E.]

Together with his colleague Josef Brix (1859–1943), who represented Town Planning in the Civil Engineering Department, Genzmer established the Town Planning, Settlement and Housing seminar during the 1907/08 Winter Semester. In 1910, both were successful in the competition for the planning of Greater Berlin. Yet neither their nor anyone else's plans chosen at the time for this large-scale project were implemented. Prior to this, in 1908, they had achieved the first and the third places in the competition for the Frohnau garden city near Berlin, which was then developed largely in line with their proposals and which even today still ranks among the parts of Berlin that are most worth seeing . Numerous other development plans show how Genzmer was concerned with every detail. For instance, he assigned substantial didactic significance to the treatment of individual town planning and architectural tasks. Of Genzmer's lesser works, the buffalo houses at the Zoological Garden are also worth mentioning.

His country house in the Podbielskiallee, in the Berlin suburb of Dahlem, is considered his finest and most mature achievement in building private houses, and it was where he also proved his talent as a designer of furniture, lighting and glazing work. In his later years he held numerous offices as a judge of competition entries and as senior consultant. In 1925, the Technische Hochschule awarded him honorary membership. Felix Genzmer died on 6 August 1929 in Dahlem.

Lit.: F. Schultze: Zum 70. Geburtstag von Felix Genzmer (On the 70th Birthday of Felix Genzmer) in: Zentralblatt der Bauverwaltung 46 (1926), p. 513; Erich Blunck: Felix Genzmer; ein Bild seines Wirkens zu seinem 70. Geburtstag (Felix Genzmer; A view of his work on his 70th Birthday) in: Deutsche Bauzeitung 60 (1926), p. 753-760.

[M. E.]

Der Rückblick auf 125 Jahre TU Berlin ist ein Blick in die Vergangenheit und kann doch zugleich mit einem Blick in die Zukunft verbunden werden: Georg Hamel wies nämlich vor einem Dreivierteljahrhundert in einer Rektoratsrede darauf hin, dass die Förderung der technischen Wissenschaften fast unmittelbar auch volkswirtschaftliche Vorteile bringe; zudem sah er die Gefahr, dass die damalige TH Berlin ohne hinreichende finanzielle Unterstützung in absehbarer Zeit hinter den anderen deutschen Hochschulen und erst recht hinter den US-amerikanischen zurückbleiben dürfte. An mahnenden Stimmen hat es somit schon frühzeitig und auch später nicht gefehlt, denn als beispielsweise nach 1975 einige Jahre lang in Deutschland die Datenverarbeitung in immer geringerem Maße gefördert wurde, nahm das nicht jeder kommentarlos hin; vielmehr organisierte *Wolfgang Giloi* ein von vielen Professoren unterzeichnetes Protestschreiben, welches zunächst zwar so gut wie wirkungslos blieb; doch als ihm später das Bundesverdienstverkreuz 1. Klasse verliehen wurde, da pries man neben seiner wissenschaftlichen Kompetenz und seinem Ideenreichtum auch seine „beharrliche Überzeugungskraft".

Wolfgang Giloi wurde am 1. Oktober 1930 in Sobernheim geboren, studierte bis 1957 an der TH Stuttgart Elektrotechnik, war anschließend dort wissenschaftlicher Mitarbeiter am Lehrstuhl für Fernmeldeanlagen und promovierte 1960 zum Dr.-Ing.; danach arbeitete er als Entwicklungsingenieur und Leiter des Rechenzentrums bei AEG-Telefunken, wurde 1965 als Professor an die TU Berlin berufen und übernahm die Leitung des Instituts für Informationsverarbeitung. Von 1971 bis 1977 lehrte er in den USA an der University of Minnesota und kehrte danach an die TU Berlin zurück, wo er bis zu seiner Emeritierung im Jahre 1996 blieb; 1983 wurde er Gründungsdirektor des Forschungsinstitutes für Rechnerarchitektur und Softwaretechnik der GMD in Berlin.

Wolfgang Giloi gehört zu denen, die schon in den Fünfzigerjahren die Bedeutung des Computers erkannten und sich damit beschäftigten,

The review of 125 years of the Technische Universität Berlin is a look back into the past – though at the same time linked to a glance towards the future. Three-quarters of a century ago, Georg Hamel pointed out in his speech as Principal that the promotion of technical sciences almost immediately resulted in macro-economic advantages. However, he also saw the danger that the (then) Technische Hochschule Berlin might well remain behind other German universities (and certainly behind American institutions) if it were not to receive sufficient financial backing in the foreseeable future. So there had not been any shortage of warning voices early on. These words of caution were also heard later on, there were some who could not remain silent after 1975, for instance, as data processing in Germany received less and less funding. Instead, *Wolfgang Giloi* organised a protest letter signed by numerous professors. This note was ineffective, though when he was later awarded the Order of Merit of the Federal Republic, 1st class, he was praised not just for his scientific competence and imaginativeness but also for his "dogged persuasiveness".

Wolfgang Giloi was born in Sobernheim on 1 October 1930, and studied electrical engineering at the Technische Hochschule in Stuttgart until 1957. He was subsequently employed as a member of scientific staff by the department for telecommunication systems and was conferred a doctorate ("Dr.-Ing.") in 1960. After that, he worked as a development engineer and head of the data processing centre at AEG-Telefunken. In 1965 he was called to the TU Berlin as professor and took charge at the Institute for Information Processing. He taught in the USA at the University of Minnesota from 1971 to 1977, before returning to the TU Berlin, where he was to remain until given emeritus status in 1996. In 1983 he became a founding director of the GMD's Research Institute for Computer Architecture and Software Technology in Berlin.

denn bereits während seiner Tätigkeit an der TH Stuttgart programmierte
er die von Konrad Zuse entwickelte Rechenanlage Z22. An der TU Berlin
baute er die erste deutsche Forschungsgruppe für Computergraphik auf
und trug entscheidend dazu bei, dass an der TU Berlin ein eigenständiger
Studiengang Informatik eingerichtet und zudem im Dezember 1970 der
Fachbereich Kybernetik gegründet wurde, der später in Informatik umbe-
nannt wurde; erwähnt sei in diesem Zusammenhang nur die wegweisende Konferenz *Der Computer in der Universität*, die im Sommer 1968
von der TU Berlin und dem MIT veranstaltet wurde und an deren Organisation Wolfgang Giloi maßgeblich beteiligt war. Jedoch besteht Wolfgang Gilois Beitrag zur Informatik vor allem in technischen Neuerungen,
die so bedeutend sind, dass er mit der Diesel-Medaille als ein herausra-
gender Erfinder ausgezeichnet worden ist; die berühmteste dieser technischen Leistungen stellt die Entwicklung des Superrechners SUPRENUM
Ende der achtziger Jahre dar.

Über seinen Forschungsschwerpunkt Rechnerarchitektur hat Wolfgang Giloi ein gleichnamiges Standardwerk verfasst, welches er Konrad
Zuse gewidmet hat, dem „Rechnerpionier"; in dessen Nachfolge ist Wolfgang Giloi ein Informatikpionier.

---

*Lit.: Simulation und Analyse stochastischer Vorgänge. München und Wien 1967.
– Rechnerarchitektur. Berlin u. a. 1993. – Über einhundert Aufsätze in Zeitschriften.*

[F. H.]

Wolfgang Giloi was one of those people who had already recognised the significance of the computer back in the nineteen-fifties, and who became involved with it, for even when he was employed at the TH Stuttgart he was programming the "Z22" computer system developed by Konrad Zuse. At the TU Berlin, he established the first German research group for computer graphics and played a major role in the establishment of informatics as an independent course of study at the TU Berlin (the Department of Cybernetics was also established in December 1970 and later renamed "Informatics"). In this connection only the pioneering *Computer in the University* conference is mentioned, held in the summer of 1968 by the TU Berlin and the MIT, and which Wolfgang Giloi was considerably involved in organising. However, Wolfgang Giloi's contribution to informatics was primarily concerned with technical innovations that were so important that he was awarded the Diesel medal for outstanding inventors – the most famous of these technical achievements being the development of the SUPRENUM supercomputer in the late nineteen-eighties.

Wolfgang Giloi wrote a standard work about computer architecture, his main area of research, a book which he dedicated to the "computer pioneer" Konrad Zuse. Wolfgang Giloi, informatics pioneer had become Zuse's successor.

Lit.: Simulation und Analyse stochastischer Vorgänge. Munich and Vienna 1967. – Rechnerarchitektur. Berlin, etc. 1993. – More than one hundred essays in periodicals.

[F. H.]

Der Sohn eines Konzertpianisten, am 2.2.1912 in Berlin geboren, absolvierte das angesehene Askanische Gymnasium und studierte ab 1930 Elektrotechnik an der TH Berlin. Nach der mit der „sehr gut" benoteten Diplomprüfung wurde er Assistent im Institut für Hochfrequenztechnik bei Heinrich Fassbender (1884-1970), der ab 1937 auch das Heinrich-Hertz-Institut (HHI) für Schwingungsforschung leitete. Bei ihm wurde *Friedrich Wilhelm Gundlach* 1938 mit Auszeichnung promoviert; mit der Dissertation begann seine Liebe zu den Laufzeiterscheinungen der Elektronen in Hochfrequenz-Bauelementen, die seine Tätigkeit in Industrie und Forschung lebenslang bestimmte.

Er trat 1938 als Laboratoriumsingenieur in die alteingesessene Berliner Firma Julius Pintsch ein, wo er bald Gruppenleiter für die Entwicklung von Laufzeitröhren wurde. Nebenamtlich unterrichtete er an der Gauss-Ingenieurschule. 1942 wechselte er zur Pintsch-Tochterfirma Funkstrahl GmbH und zog mit ihr später nach Konstanz um.

1947 habilitierte sich Gundlach an der TH Karlsruhe, wurde dort Privatdozent und arbeitete zugleich als Laborleiter des Karlsruher Werks von Siemens & Halske. Im Herbst 1949 wurde er zum Ordinarius des neugeschaffenen Lehrstuhls für Elektrotechnik und Direktor des Instituts für Fernmeldetechnische Geräte und Anlagen an die TH Darmstadt berufen. Seine „Grundlagen der Höchstfrequenztechnik" von 1949 wie auch das später mitherausgegebene „Taschenbuch für Hochfrequenztechnik" gelten als Standardwerke.

1954 wechselte Gundlach als Ordinarius und Direktor des Instituts für Hochfrequenztechnik in seine Heimatstadt und -Hochschule, die TU Berlin. Hier nahm gerade das HHI seine Arbeit wieder auf, wo er die Abteilung Hochfrequenztechnik leitete, den Forschungsschwerpunkt Laufzeitröhren aufbaute und über Kabel und Hohlleiter, Halbleiter in Mikrowellen, Digitalradar, Normalfrequenz und Maser sowie Wellenausbreitung in der

Friedrich Wilhelm Gundlach was born the son of a concert pianist on 2 February 1912. He graduated from the prestigious Askanisches Gymnasium secondary school and started to study electrical engineering at the TH Berlin in 1930. Having obtained a degree with first class honours, he became an assistant at the Institute for High-Frequency Engineering under Heinrich Fassbender (1884–1970), who in 1937 became director of the Heinrich-Hertz-Institut (HHI) for Vibration Engineering. While at this post, Gundlach earned a doctorate with distinction; his dissertation dealt with the time of flight of electrons in high-frequency components, a field of study that was to occupy him in his activities in industry and as an academic for the rest of his life.

In 1938 he became a laboratory engineer at the long-established Berlin company Julius Pintsch, where he soon became a group leader for the development of transition-time tubes. He also taught at the Gauss Ingenieurschule (engineering school). In 1942 he moved to the Pintsch subsidiary Funkstrahl GmbH and later moved with them to Konstanz.

In 1947 Gundlach gained the grade of professor at the TH Karlsruhe, where he worked as a private tutor. He was also a laboratory director at the Karlsruhe facility of Siemens & Halske. In the autumn of 1949 he was appointed to the new chair of electrical engineering and was made director of the Institut für Fernmeldetechnische Geräte und Anlagen (Institute for Telecommunication Devices and Systems) at the TH Darmstadt. His 1949 publication "Grundlagen der Höchstfrequenztechnik" (Basics of Highest-Frequency Technology) and his "Taschenbuch der Hochfrequenztechnik" (Handbook of High-Frequency Technology) are regarded as standard works.

In 1954 Gundlach returned to his native Berlin and to his alma mater, the TU, where he took the chair and directorship of the Institute of High-Frequency Technology. The HHI, where he was director of the High-

Atmosphäre arbeitete. Unter seiner Leitung konnte 1968 das neue Institutsgebäude eröffnet werden. Er hatte maßgeblich Anteil daran, dass das HHI 1975 in eine Bund-Land-GmbH umgewandelt wurde.

An der TU erforschte und lehrte er unterschiedlichste Gebiete auf der gemeinsamen Grundlage der Hochfrequenztechnik: Für die Mikrowellenröhrentechnik wurden neue Verfahren und Werkstoffe entwickelt, neue Methoden der Kernstrahlenmessung ermöglichten eine genauere Einschätzung der Gefährlichkeit, mit einem TV-Farbumkehrverfahren konnte die Kopie vom Negativ sofort reglergesteuert gesendet werden. Als Rektor der TU 1965–1967 begegnete er mit seiner ausgleichenden, sachlichen und zugleich entschiedenen Haltung den Spannungen des aufkeimenden Generationenkonflikts. 1979 emeritiert, würdigte die TU 1987 seine Verdienste als Mitglied des Kuratoriums, der Hauptkommission und der Stellenkommission sowie als Leiter der Forschungskommission des Fachbereichs Elektrotechnik mit der Verleihung der Ehrensenatorwürde. Für seine vielfältige Verbands-, Forschungs- und Lehrtätigkeit auf dem Gebiet der Mikrowellentechnik verlieh ihm der VDE den Ehrenring, für seine bahnbrechenden Arbeiten in der Hochfrequenztechnik wurde er mit dem Bundesverdienstkreuz und der Ernennung zum Fellow des Institute of Electrical and Electronics Engineers geehrt. Gundlach starb am 27.1.1994 in Berlin.

Lit.: TU-Archiv

[B. E.]

Frequency Department, once again took up his research, focussing on transition-time tubes. He also worked on cable and hollow conductors, semiconductors in microwaves, digital radar, standard frequency and masers, as well as wave transmission in the atmosphere. A new building for the institute was opened in 1968. It was partly due to Gundlach's efforts that the HHI became a limited company under the auspices of the federal and Berlin state governments.

At the TU he taught and researched a very wide range of subjects in the field of high-frequency technology: new processes and materials were developed for microwave tube technology, new methods of measuring nuclear radiation made it possible to make more precise assessments of danger, and a process that reversed TV colours meant the copy of a negative image could be broadcasted instantly by a regulator. As rector of the TU in 1965–1967 he dealt with the tensions of the burgeoning generational conflict with his balanced, matter-of-fact and decisive approach. He became professor emeritus in 1979 and the TU awarded him by making him an honorary senator for his services as a member of the Kuratorium, of the High Commission and of the Stellenkommission as well as director of the research committee for electrical engineering. The German Association of Electrical Engineers awarded Gundlach with the ring of honour for his wide range of institutional, research and teaching activities in the field of microwave technology. The German state admitted him to the Order of Merit of the Federal Republic for his groundbreaking work on high-frequency technology and he was also made a fellow of the Institute of Electrical and Electronics Engineers. Gundlach died in Berlin on 27 January 1994.

Lit.: TU-Archive

[B. E.]

Das elektronische Rechengerät werde zum Rechenzentrum der industriellen Forschung werden, und auch Berlin benötige ein solches Gerät, damit die TU nicht zu einer Hochschule zweiten Ranges werde, bemerkte im Jahre 1953 weitsichtig eine von *Wolfgang Haack* gegründete Arbeitsgruppe der TU Berlin.

Wolfgang Haack wurde am 24. April 1902 in Gotha geboren, studierte erst in Hannover Maschinenbau, dann in Jena Mathematik und promovierte dort 1926 mit der Schrift *Die Bestimmung von Flächen, deren geodätische Linien durch die Abbildung in die (x;y) Ebene (durch x=u; y=v) in Kegelschnitte übergehen.* Danach studierte er noch in Hamburg, bevor er als Assistent an die TH Stuttgart ging und sich 1929 an der TH Danzig mit einer Arbeit über affine Differentialgeometrie habilitierte. 1935 wechselte er an die TH Berlin und folgte 1937 einem Ruf an die TH Karlsruhe. Die TH Berlin berief ihn zwar 1944, doch Wolfgang Haack konnte diese Aufgabe infolge der Kriegsumstände nicht mehr wahrnehmen; 1949 wurde er als Nachfolger Georg Hamels endgültig von der TU Berlin zum Professor der Mathematik und Mechanik ernannt. 1964 wurde Wolfgang Haack auf den neuen Lehrstuhl für numerische Mathematik berufen, den er bis zu seiner Emeritierung im Jahre 1968 innehatte.

Die beiden Lehrstühle deuten bereits an, dass Wolfgang Haack sich auf mannigfaltigen Gebieten ausgezeichnet hat. Als Mathematiker beschäftigte er sich anfangs mit der Geometrie, insbesondere mit der Differentialgeometrie, bevor er sich seit 1936 technischen Fragestellungen zuwandte und beispielsweise Arbeiten über Gasdynamik und Differentialgleichungen verfasste. Auch später an der TU Berlin hat er sich um technische Probleme gekümmert und mehrere Aufsätze über die Automatisierung der Flugsicherung geschrieben. Seine bedeutendsten mathematischen Forschungen während der Zeit an der TU galten der Theorie der partiellen Differentialgleichungen, wozu er grundlegende Ergebnisse beisteuerte; überdies gelang es ihm, den wissenschaftlichen Nachwuchs

The electronic computing device was becoming the primary means of calculation in industrial research, and Berlin also needed such a machine to prevent the Technische Universität from deteriorating to a second class academic institution, noted a far-sighted task force at the Technische Universität Berlin, set up by *Wolfgang Haack* in 1953.

Wolfgang Haack was born in Gotha on 24 April 1902, and first studied mechanical engineering in Hanover, followed by Mathematics in Jena, where he was awarded a doctorate in 1926 for the thesis *Die Bestimmung von Flächen, deren geodätische Linien durch die Abbildung in die (x;y) Ebene (durch x=u; y=v) in Kegelschnitte übergehen* (The determination of areas whose geodetic lines become conic sections when depicted in the (x;y) plane (through x=u; y=v). He then studied in Hamburg, before going to the TH Stuttgart as an assistant and qualified as a university lecturer at the TH Danzig (now Gdansk) in 1929 with a paper on affine differential geometry. In 1935 he moved to the TH Berlin before following a call to the TH Karlsruhe in 1937. Although the TH Berlin did invite him to work there in 1944, Wolfgang Haack was unable to take up the post because of the war. In 1949 he was finally named by the TU Berlin as the successor to Georg Hamel as Professor of Mathematics and Mechanics. In 1964 Haack was called to the new chair for numerical mathematics, a position he was to hold until being given emeritus status in 1968.

These two chairs showed that Haack had distinguished himself in a variety of fields. As a mathematician, he had initially concerned himself with geometry, in particular with differential geometry, before he turned to technical questions in 1936, publishing works on gas dynamics and differential equations, for example. Later at the Technische Universität Berlin, he was also involved with technical problems, writing a number of papers on the automation of air traffic control. His most important mathematical research during his time at the TU dealt with the theory

zu fruchtbaren Beiträgen zu ermuntern, sodass in jenen Jahren an der TU Berlin einige Habilitationsschriften und ein rundes Dutzend Dissertationen über Differentialgleichungen entstanden.

Doch nicht nur dem Forscher und Hochschullehrer Wolfgang Haack hat die TU Berlin viel zu verdanken, sondern auch dem Organisator, denn mit seinem Namen ist untrennbar die Einführung des Computers an der TU verbunden. Als er nach dem Kriege von den elektronischen Rechnern erfahren hatte, erfasste er deren Bedeutung hinsichtlich der zukünftigen Wissenschaft und bildete 1950 eine Arbeitsgemeinschaft für elektronische Rechengeräte. Wolfgang Haack bemühte sich in den nächsten Jahren, an der TU Berlin einen Rechner aufstellen zu lassen, und nahm Kontakt zu Konrad Zuse auf. Das größte Hindernis bildete die Finanzierung; so erhielt Haack von der Deutschen Forschungsgemeinschaft eine abschlägige Antwort, weil es laut DFG für die deutschen Universitäten vollkommen ausreichend sei, wenn Göttingen, Darmstadt und München auf dem Gebiet der elektronischen Rechner arbeiteten. Daraufhin sammelte er bei verschiedenen Firmen Spenden und bekam schließlich die erforderliche Summe zusammen, sodass 1958 vor allem dank Wolfgang Haack der erste Computer an der TU Berlin in Betrieb gehen konnte.

Lit.: Differentialgeometrie. 2 Bände. Wolfenbüttel und Hannover 1948. – Elementare Differentialgeometrie. Basel und Stuttgart 1955. – Vorlesungen über Partielle und Pfaffsche Differentialgleichungen (mit Wolfgang Wendland). Basel und Stuttgart 1969.

[F. H.]

of partial differential equations, to which he contributed a number of fundamental findings. Furthermore, he succeeded in motivating young scientists to make productive contributions, so that a number of postdoctoral theses and around a dozen dissertations were written on the subject of differential equations at the TU Berlin during those years.

However, the TU Berlin not only had much to thank the researcher and lecturer Wolfgang Haack for, but it also had to acknowledge him as an organiser, for his name is inseparably linked with the introduction of the computer to the Technische Universität. When, after the war, he found out more about electronic computers, he was seized by their future importance for science, and he founded a working group for electronic computing devices in 1950. During the years to come, Haack tried to have a computer installed at the TU Berlin and to this end contacted Konrad Zuse. The biggest handicap was finance, and Haack was turned down by the *Deutsche Forschungsgemeinschaft* (German Research Council) as this body considered it perfectly satisfactory for Germany's Universities that Göttingen, Darmstadt and Munich were actively working in the area of electronic computers. In response he managed to obtain donations from a number of companies, finally acquiring the funds needed so that the TU Berlin's first computer could begin work in 1958, thanks above all to his work.

Lit.: Differentialgeometrie. 2 Volumes. Wolfenbüttel and Hanover 1948. – Elementare Differentialgeometrie. Basel and Stuttgart 1955. – Vorlesungen über Partielle und Pfaffsche Differentialgleichungen (with Wolfgang Wendland). Basel and Stuttgart 1969.

[F. H.]

Robert von Halász (*1905)

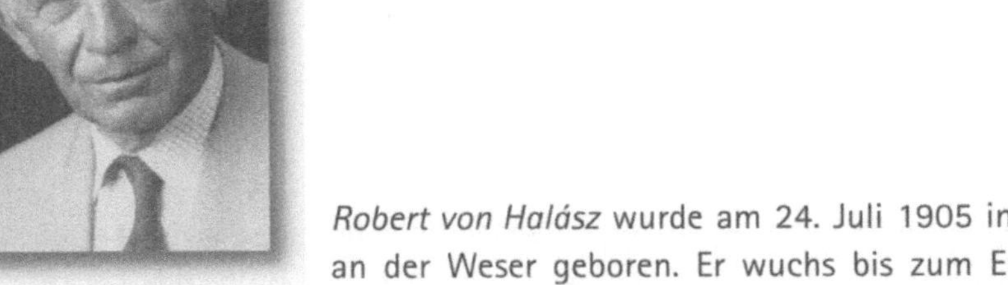

Robert von Halász wurde am 24. Juli 1905 in Höxter an der Weser geboren. Er wuchs bis zum Ende des 1. Weltkrieges in Colmar auf, dann wurde die Familie aus dem Elsass ausgewiesen, da der Vater dort als deutscher Beamter tätig gewesen war. Die Familie zog nach Berlin, wo von Halász seine Abiturprüfung absolvierte. Von 1925 bis 1930 studierte er Bauingenieurwesen an der TH Berlin.

Da von Halász nach Abschluss seines Studiums keine Anstellung als Bauingenieur fand, übernahm er zunächst die Betriebsleitung und Geschäftsführung der Formsand- und Braunkohlegruben Petersburg. Von 1936 bis 1942 war er dann als Chefkonstrukteur und Leiter des Technischen Büros bei der Firma A. Plattner KG in Berlin angestellt. Hier war er für die Planung von Holzbauten aus serienmäßig vorgefertigten Bauelementen zuständig. Ab 1937 unterrichtete von Halász nebenamtlich als Dozent an der Ingenieurschule Berlin und war als Referent in der Reichsstelle für Baustatik tätig.

Aus seinen Erfahrungen mit dem Holzbau entwickelte von Halász die Idee, Hochbauten aus Stahlbetonfertigteilen zu errichten, und fand bei der Preußischen Bergwerks- und Hütten-AG (Preussag) potente Unterstützung. Die Preussag errichtete 1941 in Rüdersdorf ein Betonwerk, um die Serienfabrikation von Industriehallen aus Stahlbetonelementen aufzunehmen. Von Halász wurde von seinen Aufgaben bei der Reichsstelle für Baustatik freigestellt und zur Preussag abgeordnet – die Preussag war zur Hälfte im Staatsbesitz. 1943 wurde von Halász Chefingenieur und Leiter des Rüdersdorfer Betonwerks der Preussag. Nach umfangreichen Forschungsarbeiten theoretischer, konstruktiver, aber auch marktanalytischer Art realisierte er hier zum ersten Mal die serienmäßige Herstellung kompletter Industriehallen mit Kranbahn aus vorgefertigten Stahlbetonbauteilen, die ab Lager verkauft wurden.

Robert von Halász was born in Höxter an der Weser on 24 July 1905. Until the end of the 1st World War he was brought up in Colmar, then his family was expelled from the Alsace region because his father had been active as a German civil servant there. The family moved to Berlin, where von Halász sat his *Abitur* examinations. From 1925 to 1930 he studied civil engineering at the TH Berlin.

Since von Halász was unable to find employment as a civil engineer on completing his studies, he first took over the production management and direction of the Formsand- und Braunkohlegruben Petersburg (Petersburg moulding sand and lignite mines) company. From 1936 to 1942 he was employed as chief designer and head of the technical office at A. Plattner KG in Berlin, where he was responsible for planning timber buildings from prefabricated components. From 1937 onward he also lectured part-time at the School of Engineering in Berlin and was active as an expert at the Office for Structural Analysis.

From his own experience with timber construction, von Halász developed the idea of erecting above-ground structures from precast reinforced concrete components and he found keen support from the Preußische Bergwerks- und Hütten-AG (Prussian Mine and Ironworks AG), better known as Preussag. The company founded a concrete works in Rüdersdorf in 1941 in order to start the mass production of factory buildings from reinforced concrete components. Von Halász was granted release from his duties at the Office for Structural Analysis and was delegated to Preussag, a company that was then half-owned by the state. In 1943, von Halász became chief engineer and head of Preussag's Rüdersdorfer concrete works. After extensive research work of a theoretical, constructional and market-analysis nature, he was able to accomplish the very first series production of complete factory buildings (with crane track) from prefabricated reinforced concrete components, sold from stock.

1946 erhielt von Halász einen Lehrauftrag an der gerade gegründeten Technischen Universität Berlin, 1948 folgt der Ruf auf den Lehrstuhl für Baukonstruktion. Hier konnte er seine Ideen eines ganzheitlich technischen Bauentwurfs und einer Industrialisierung des Bauens an seine Studenten vermitteln. Er reformierte die Vorlesung über Baukonstruktion und führte die Bauphysik ins Studium ein, womit er die Ausbildung der Bauingenieure auf einen zeitgemäßen Stand brachte. Von Halász blieb jedoch weiterhin auch der Praxis verbunden. Seine Idee der Fertigbauweise aus Stahlbeton erreichte bald den Wohnungsbau, und er selbst leistete grundlegende Beiträge zur Entwicklung des Großtafelbaus. Auch auf dem Gebiet der Holzkonstruktionen hat er weitere richtungsweisende Arbeiten publiziert. Seine Erfahrungen im Ingenieurholzbau flossen in diverse Beiträge im „Holzbautaschenbuch" ein, das von Halász ab 1943 herausgab. Außerdem war er von 1954 bis 1974 Schriftleiter der Zeitschrift „Bautechnik".

Nach seiner Emeritierung 1973 gründete von Halász mit jüngeren Kollegen ein Ingenieurbüro in Berlin und gab seine vielfältigen Erfahrungen nun als Planer, Prüfingenieur und Gutachter weiter. Robert von Halász erhielt 1980 das Bundesverdienstkreuz und 1982 die Ehrendoktorwürde der Universität Dortmund verliehen.

Lit.: Klaus Stiglat: Robert von Halász, in ders. (Hg.): Sie bauen und forschen. Bauingenieure und ihr Werk (Beton- und Stahlbetonbau Spezial), Berlin 1999, S. 36–40

[J. Z.]

In 1946 he was appointed to teach at the recently founded Technische Universität Berlin, and in 1948 heeded the call to the Chair for Building Construction. Here he was able to put across to his students his ideas of integrated design in technical building, and of the industrialization of the building trade. He reformed the teaching of building construction and ensured that building physics became one of the subjects studied, so modernizing the training of civil engineers. However, von Halász remained a practical man, too. His idea of assembling prefabricated buildings made of reinforced concrete soon caught on in the housing construction sector, and he made a number of important contributions to the development of the large panel construction method. He also published a number of pioneering works on the subject of timber building methods. His experience of timber construction found its way into a variety of contributions to the "Holzbautaschenbuch" (timber construction manual) that von Halász started publishing in 1943. In addition, he edited the "Bautechnik" (building technology) magazine from 1954 to 1974.

After his retirement in 1973, von Halász founded an engineer's office in Berlin together with a number of younger colleagues. There he was able to pass on his great experience as planner, test engineer and consultant. Robert von Halász received the Bundesverdienstkreuz (Order of Merit) in 1980 and in 1982 he was awarded an honorary doctorate from the University of Dortmund.

Lit.: Klaus Stiglat: Robert von Halász, in the same (ed.): Sie bauen und forschen. Bauingenieure und ihr Werk (Beton- und Stahlbetonbau Spezial). Berlin 1999, p. 36–40

[J. Z.]

Wer nichts als Chemie verstehe, verstehe auch die nicht recht, sprach einst Lichtenberg, und insofern lässt sich mit Fug und Recht behaupten, dass *Georg Hamel* weit mehr als nur Mechanik verstanden hat.

Georg Hamel wurde am 12. September 1877 in Düren geboren und studierte von 1895 bis 1897 an der TH Aachen Mathematik, Physik und „einige grundlegende Fächer der technischen Wissenschaften"; von 1897 bis 1900 an der Berliner und 1900/1901 an der Göttinger Universität studierte er jeweils Mathematik, Physik und Philosophie und promovierte 1901 bei David Hilbert höchstselbst mit der Arbeit *Ueber die Geometrieen, in denen die Graden die Kürzesten sind.* 1903 habilitierte sich Hamel in Karlsruhe für Mathematik und Mechanik, 1905 wurde er von der TH Brünn zum Professor der Mechanik berufen, 1912 von der TH Aachen zum Professor der Mathematik, und seit 1919 lehrte er an der TH Berlin als Professor der Mathematik und der Mechanik. 1933 vollzog er mit irritierendem und übersteigertem, geradezu skurrilem Pathos eine Selbstgleichschaltung; jedoch stimmt es nicht, dass auch er den Aufbau einer „Deutschen Mathematik" gefordert habe. Gegen Kriegsende wurde er ausgebombt, verließ Berlin und kehrte später nur kurz zurück, um 1948/49 acht Wochen lang an der Humboldt-Universität vorzulesen. An der TU Berlin hat er keine Vorlesungen mehr gehalten und wurde 1949 emeritiert.

Georg Hamel hob hervor, dass er David Hilbert eine besonders reiche wissenschaftliche Anregung verdanke; in einer Vorrede gedachte er Immanuel Kants als seines Lehrers – und in keinem der beiden Fälle kokettiert er mit einem berühmten Namen, denn Hamels wissenschaftliches Werk ist wesentlich durch Kant und Hilbert geprägt. Hervorgetan hat er sich in der Mechanik, der Philosophie und der Mathematik; hierin verfasste er Schriften über Funktionentheorie, Variationsrechnung, Differential- und Integralgleichungen, Zahlentheorie und – wie mit seiner Dissertation – über Geometrie. Schon diese offenbart seine philosophischen

A man who understands nothing but chemistry will
not even comprehend that properly, Georg Christoph
Lichtenberg once said, so we can claim with complete justification that
Georg Hamel understood far more than just mechanics.

Georg Hamel was born in Düren on 12 September 1877 and studied
mathematics, physics and "a number of fundamental areas of technical
sciences" at the TH Aachen from 1895 to 1897. Then, from 1897 to 1900
at the Berlin University and in 1900/1901 at the University of Göttingen, he studied mathematics, physics and philosophy, being conferred his
doctorate by none less than David Hilbert in person with the thesis *Ueber
die Geometrieen, in denen die Graden die Kürzesten sind* (On geometri-
cal forms in which the straight lines are the shortest) in 1901. In 1903,
Hamel qualified as a university lecturer for Mathematics and Mechanics
in Karlsruhe, and was appointed Professor of Mechanics by the Technische Hochschule Brünn (now Brno, Czech Republic) in 1905. In 1912
he was appointed Professor of Mathematics by the TH Aachen, teaching
at the Technische Hochschule Berlin as Professor of Mathematics and
Mechanics from 1919. In 1933, when the Nazis took power, he subju-
gated himself to them – with an irritating and exaggerated, downright
absurd pathos. Nevertheless, it is not true that he had also demanded the
establishment of "Germanic Mathematics". Towards the end of the War
he was bombed out, left Berlin and only briefly returned to lecture at the
Humboldt Universität – for eight weeks in 1948/49. He gave no more
lectures at the TU Berlin and was granted emeritus status in 1949.

Hamel emphasized that he had received much scientific stimu-
lus from David Hilbert. In an introductory speech, he commemorated
Immanuel Kant as his mentor – in neither case was he indulging in mere
name-dropping, for Hamel's scientific work was indeed greatly influenced by Kant and Hilbert. He distinguished himself in mechanics, philosophy and mathematics, and in these fields wrote papers on the theory

Neigungen, denn er widmet sich der grundsätzlichen Frage nach der allgemeinsten Geometrie der Art, dass darin die Gerade die kürzeste Linie sei; gelungen ist Hamel eine auch nach über einhundert Jahren noch fesselnde Antwort.

Jene Arbeit stellt zugleich einen Versuch dar, um sich in den axiomatischen Aufbau eines Gedankengebäudes einzuüben, und das wird Georg Hamels berühmtester Beitrag zur Mechanik und zur Philosophie werden. Wissenschafts- und auch erkenntnistheoretisch ist er unübersehbar bei Kant und dessen transzendentalem Idealismus in die Schule gegangen; so steht es außer Zweifel, dass Raum, Zeit und Kraft apriorische Formen sind. Hamel möchte wie sein großes Vorbild Lagrange in der *Mécanique analytique* analytisch vorgehen; das bedeutet die Verwendung einer deduktiven Methode, „die das ganze Gebäude einer Wissenschaft aus wenigen grundlegenden Prinzipien zu errichten strebt." Danach wird die klassische Mechanik aus diesen Axiomen auf rein logischem Wege abgeleitet.

Entstanden sind die *Elementare Mechanik* und als krönender Abschluss die *Theoretische Mechanik* – zwei philosophische Werke über mathematische Mechanik, die sich mit einem Worte Goethes über Hamels Lehrer Kant würdigen lassen: Beim Lesen wird einem zumute, als träte man in ein helles Zimmer.

Lit.: Ueber die Geometrieen, in denen die Graden die Kürzesten sind. Göttingen 1901. – Über Raum, Zeit und Kraft als apriorische Formen der Mechanik. In: Jahresbericht der DMV 18. 1909, S. 357–385. – Elementare Mechanik. Leipzig 1912. – Theoretische Mechanik. Berlin u. a. 1949.

[F. H.]

of functions, the calculus of variations, differential and integral equations, number theory and (as in his dissertation) about geometry. This in itself revealed his philosophical inclinations as he devoted himself to the fundamental question dealing with the most basic geometry of the kind where the straight line is the shortest. Hamel succeeded in finding an answer, which over a hundred years later, is still enthralling.

At the same time, this work also represented an attempt to practise the axiomatic construction of an edifice of ideas that was to become Georg Hamel's most famous contribution to mechanics and philosophy. In terms of scientific theory and theory of knowledge he was clearly a follower of the school of Immanuel Kant and his transcendental idealism – thus there was no doubt that space, time and energy were a priori parameters. Hamel wanted to proceed analytically like his great role-model Lagrange in Mécanique analytique, which meant using a deductive method where "the whole structure of a science was to be erected from a few fundamental principles". In accordance, classical mechanics is derived from these axioms in a purely logical way.

The results were Elementare Mechanik (Elementary mechanics) and its culmination Theoretische Mechanik (Theoretical mechanics), two philosophical works about mathematical mechanics that can be acknowledged with a phrase from Goethe used to describe Hamel's mentor Kant: "when reading it, one has the sensation of entering a well-lit room".

Lit.: Ueber die Geometrieen, in denen die Graden die Kürzesten sind. Göttingen 1901. – Über Raum, Zeit und Kraft als apriorische Formen der Mechanik. In: Jahresbericht der DMV 18. 1909, p. 357–385. – Elementare Mechanik. Leipzig 1912. – Theoretische Mechanik. Berlin, etc. 1949.

[F. H.]

Herta Hammerbacher (1900–1985)

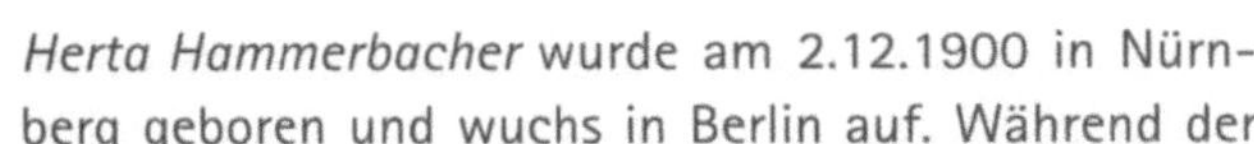

Herta Hammerbacher wurde am 2.12.1900 in Nürnberg geboren und wuchs in Berlin auf. Während der Gärtnerlehre 1917–19 in den Schlossgärten von Sanssouci lernte sie den Staudenzüchter Karl Foerster (1874–1970) kennen. Nach kurzer Berufspraxis am Bodensee, Studium an der Lehr- und Forschungsanstalt für Gartenbau in Berlin-Dahlem 1924–26 und Tätigkeit in der Abteilung Gartengestaltung der Baumschule L. Späth in Berlin-Treptow schloss sie sich 1928 der Arbeitsgemeinschaft Karl Foerster und Hermann Mattern (1902–1971) an, die bis 1948 bestand. Mit Mattern war sie von 1928 bis 1935 verheiratet, das Paar hatte zwei Töchter.

Hammerbacher wollte den landschaftsgebundenen Garten verwirklichen. Ökologisches Bewusstsein und Umweltschonung bestimmten ihre Gestaltungsprinzipien schon zu einer Zeit, als diese längst noch nicht in der Öffentlichkeit anerkannt waren. Der Wohngarten war eines ihrer Ziele. Damit prägte sie maßgeblich den Stil der Freiraumgestaltung vor allem der Fünfziger- und Sechzigerjahre. Etwa 3500 von ihr allein oder gemeinschaftlich realisierte private und öffentliche Projekte bilden ihr umfangreiches Lebenswerk. Für Berlin seien Anlagen im Waldfriedhof Zehlendorf, in der Umgebung des ehemaligen Hilton-Hotels, im Hansaviertel-Nord, in der Otto-Suhr-Siedlung, auf dem TU-Nordgelände oder im Sommergarten am Funkturm genannt, manches davon längst wieder umgestaltet.

Schon früh kooperierte Hammerbacher mit namhaften Architekten, unter ihnen Hans Scharoun (1893–1972), der nach dem Krieg Baustadtrat war und sie 1946 der TU als Lehrbeauftragte für Landschafts- und Gartengestaltung empfahl. 1950 war sie dort a.o. und von 1962 bis 1969 o. Professorin. Als Landschaftsgestalterin und –planerin konnte sie dabei eine große Wirkung erzielen. Am Herzen lag ihr aber auch das Thema „Frau im Beruf", und von ihren Studentinnen verlangte sie nach ihrem Beispiel, sich auf ein Leben mit Familie und Beruf einzustellen.

Herta Hammerbacher was born on 2 December 1900 in Nuremberg and raised in Berlin. While learning horticulture between 1917 and 1919 in the gardens of Sanssouci Palace in Potsdam, she became acquainted with Karl Foerster (1874–1970), a breeder of herbaceous perennials. She worked for a short period at Lake Constance, and then studied at the Lehr- und Forschungsanstalt für Gartenbau (Teaching and Research Institute for Horticulture) in Dahlem, Berlin between 1924 and 1926. After that, she worked in the landscape gardening department of the Ludwig Späth nursery in Treptow, Berlin. In 1928 she decided to join Karl Foerster and Hermann Mattern (1902–1971) in a working team that existed until 1948. She was married to Mattern from 1928 to 1935 and they had two daughters.

Hammerbacher wanted to realise a garden that was integrated with its landscape. Her design principles were determined by ecological awareness and environmental protection at a time when these concepts were barely heard of in general discourse. The garden for living was one of her goals. She thereby had a significant influence particularly on the style of open space design of the 1950s and 1960s. Her extensive life's work is reflected in around 3,500 private and public projects that she carried out either alone or as a collaborator. Notable Hammerbacher designs in Berlin include the Waldfriedhof Zehlendorf (Zehlendorf Forest Cemetery), the surroundings of the former Hilton Hotel, the northern part of the Hansaviertel district, the Otto-Suhr-Siedlung, the northern part of the TU grounds and the summer garden of the Funkturm (radio tower), although some of these creations have since been redesigned.

Hammerbacher worked from an early stage with renowned architects, including Hans Scharoun (1893–1972), who was city councillor for buildings after World War II. He recommended her to the post of teacher of landscape and garden design at the TU in 1946. She was an assistant professor from 1950 and a full professor from 1962 to 1969. As

Zusammen mit ihrer Tochter Merete Mattern engagierte sie sich nach der Emeritierung in der „Gesellschaft für Experimentelle und Angewandte Ökologie", einem interdisziplinären Gesprächs- und Arbeitskreis; gleichzeitig war sie Gründungsmitglied der Karl-Foerster-Stiftung für angewandte Vegetationskunde und pflegte das Erbe des Bornimer Kreises. Sie starb am 25.5.1985 in Niederpöcking bei Starnberg.

Herta Hammerbacher erfährt auch heute noch in Fachkreisen große Wertschätzung. Zu den zehn als Gartendenkmal unter Schutz gestellten Anlagen zählen die Außenanlagen des Architekturgebäudes der TU sowie acht meist in den Dreißigerjahren entstandene Hausgärten. Ihr Nachlass bildet die Basis eines DFG-geförderten Forschungsprojekts an der TU.

Lit: Beiträge zur Problematik zwischen Freiraum und Bauwerk, Festschrift Herta Hammerbacher ... zum 75. Geburtstag, Berlin 1975; Reitsam, Charlotte: Herta Hammerbacher, in: Garten und Landschaft 108 (1998) H. 11, S. 38–41; Richard, Winfried: Herta Hammerbacher zum 100. Geburtstag, in: Architektur in Berlin: Jahrbuch 8 (2000), S. 158–161; Poblotzki, Ursula: Herta Hammerbacher – eigenwillig und erfolgreich, in: Garten und Landschaft 111 (2001) H. 3, S. 36–38; TU-Archiv.

[M. E.]

a practising landscape designer and planner she had a strong impact at the university. The status of professional women was a subject close to her heart and she presented herself as an example when she demanded of her female students that they prepare themselves for a life combining family and a career.

After becoming professor emeritus in 1969, Hammerbacher and her daughter Merete Mattern devoted themselves to the Gesellschaft für Experimentelle und Angewandte Ökologie (Society for Experimental and Applied Ecology), which was an interdisciplinary cooperative and discussion forum; at the same time Hammerbacher was a founding member of the Karl-Foerster-Stiftung für angewandte Vegetationskunde (Karl Foerster Foundation for Applied Vegetation Science) and she cared for the endowment of the Bornimer Kreis. She died near Starnberg on 25 May 1985.

Herta Hammerbacher is still regarded very highly in her field today. Among the ten gardens of hers that have been listed as preserved sites are the exterior layout of the TU architecture building and eight private gardens, most of them designed during the 1930s. Her estate forms the basis of a research project at the Technical University that receives aid from the Deutsche Forschungsgemeinschaft (German Research Foundation).

Lit.: Beiträge zur Problematik zwischen Freiraum und Bauwerk, Festschrift Herta Hammerbacher ... zum 75. Geburtstag, Berlin 1975; Reitsam, Charlotte: Herta Hammerbacher, in: Garten und Landschaft 108 (1998) vol. 11, p. 38-41; Richard, Winfried: Herta Hammerbacher zum 100. Geburtstag, in: Architektur in Berlin: Jahrbuch 8 (2000) p. 158-161; Poblotzki, Ursula: Herta Hammerbacher – eigenwillig und erfolgreich, in: Garten und Landschaft 111 (2001) vol. 3, p. 36-38; TU-Archive.

[M. E.]

Hans Hausner (*1927)

Hans Hausner wurde am 23. Mai 1927 in Neustadt/ Waldnaab geboren, studierte von 1946 bis 1951 Chemie an der TU München und promovierte dort anschließend über *Spektralanalytische methodische Untersuchungen im Gleich- und Wechselstrombogen.* Danach arbeitete er von 1954 bis 1961 in der Industrie; zunächst leitete er in einem Betrieb die Forschungs- und Entwicklungsabteilung im Bereich der Elektrokeramik, war entscheidend an der Entwicklung der Titanate und Ferrite beteiligt und wandte sich 1957 der Erforschung feuerfester Materialien für Glasschmelzen zu.

1961 wechselte Hans Hausner zur Europäischen Atomgemeinschaft nach Brüssel, um die Nuklearkeramik zu erforschen; erst erfolgte ein vierjähriger Gastaufenthalt in Kalifornien beim General Electric Nuclear Research Lab, bevor er dann von 1965 beim Forschungszentrum der EURATOM im italienischen Ispra tätig wurde; dort wurden anspruchsvolle Kenntnisse verlangt, denn es galt zu erforschen, wie sich Uranoxid beim Einsatz in Brennelementen für Wasserreaktoren oder wie sich Nuklearbrennstoffe für Hochtemperaturreaktoren verhalten.

1972 gelang es der TU Berlin, Hans Hausner als Professor der nichtmetallischen anorganischen Werkstoffe zu berufen, und er lehrte dort bis zu seiner Emeritierung im Jahre 1994. Die vor Antritt seiner Professur sowohl in der Industrie als auch in Forschungseinrichtungen gesammelten Erfahrungen wirkten sich ungemein positiv auf seine Lehrtätigkeit aus, wie seine Schüler zu berichten wissen; zudem legte er großen Wert auf die Zusammenarbeit zwischen Industrie und Universität, zwischen dem Praktiker im Betrieb und dem Wissenschaftler, weil es das Ziel jeglicher technisch-wissenschaftlichen Entwicklung sei, diese zur „industriellen Reife" zu führen.

Auch während seiner Zeit an der TU Berlin blieb Hans Hausner der Forschung verbunden, veröffentlichte zahlreiche Artikel vornehmlich

Hans Hausner was born in Neustadt/Waldnaab on 23 May 1927, studied chemistry at the Technische Universität Munich from 1946 to 1951, being awarded a doctorate there for his work on spectral-analytical methodic investigations in direct and alternating current arcs. From 1954 to 1961 he worked in industry: first of all he ran the research and development department at a company manufacturing electroceramics, then he played a decisive role in the development of titanates and ferrites, before turning his attention to the exploration of fireproof materials for glass melting in 1957.

In 1961, Hausner moved to the European Atomic Energy Community (EURATOM) in Brussels to research into nuclear ceramics, followed by a four-year stay in California at the General Electric Nuclear Research Lab before recommencing work for EURATOM at its research center in Ispra, Italy in 1965. This demanding assignment called for special expertise as it was necessary to find out about the behaviour of uranium oxide in the fuel elements for water reactors and of the nuclear fuels needed for high-temperature reactors.

In 1972 the Technische Universität Berlin succeeded in appointing Hans Hausner as professor for non-metallic inorganic materials, and he taught there until he was granted emeritus status in 1994. The experience he had gained in industry and in research before taking up his professorship had an exceptionally positive effect on his teaching as his students can report. In addition, he placed much emphasis on cooperation between industry and university as well as between practical workers and scientists, because it was the goal of any technical and scientific development to bring it to "industrial maturity".

Even during his time at the Technische Universität Berlin, Hausner remained tied to research, above all publishing numerous items about ceramics, over the years becoming one of the leading experts in this field

über Keramik und stieg im Laufe der Jahre in diesem Bereich zu einem der weltweit bekanntesten Wissenschaftler auf; zu seinen Forschungsschwerpunkten gehören die Eigenschaften und das Sinterverhalten keramischer Pulver, die Anwendung der Gefriertrocknung bei der Herstellung und Formgebung der Keramik und ganz allgemein die Hochleistungskeramik. Doch nicht nur als Forscher und Hochschullehrer erwarb sich Hans Hausner einen guten Namen, vielmehr übernahm er auch Verantwortung in Organisationen wie der Deutschen Keramischen Gesellschaft, deren Vorsitzender er von 1987 bis 1991 war; in dieser Funktion betonte er, dass es für ihn nur eine Keramik gebe, jedoch nicht eine der klassischen und eine der modernen Werkstoffe, und er bemühte sich um eine Überwindung der vermeintlichen Gegensätze. Besonders hervorgehoben sei noch, dass er die Notwendigkeit einer internationalen Zusammenarbeit sah und wesentlich zur Gründung der Europäischen Keramischen Gesellschaft beigetragen hat, welcher er überdies von 1989 bis 1991 vorstand.

Schüler und Kollegen rühmen Hans Hausners menschliche Seite, als Forscher genießt er ein hohes Ansehen, weshalb ihm viele nationale und internationale Ehrungen zuteil geworden sind und es in einem Tagungsbericht über ihn heißt, dass die Hörsäle gefüllt seien, wenn Hans Hausner an das Rednerpult trete: „Und niemals verlässt man den Raum ohne Erkenntnis von bleibendem Wert."

Lit.: Über einhundert Veröffentlichungen in Sammelbänden, Konferenzberichten und Zeitschriften wie Ceramic forum international.

[F. H.]

worldwide – the main thrusts of his research included the properties and sintering behaviour of ceramic powders, the use of freeze-drying in manufacturing and forming ceramics, and high-performance ceramics in general. Yet it was not only as a researcher and university lecturer that Hausner gained a good reputation. Indeed, he also took on responsibility in organisations such as the *Deutsche Keramische Gesellschaft* (German Ceramics Society) of which he was chairperson from 1987 to 1991. In this function he emphasized that there was only one study of ceramics for him – not one study of classical and one of modern materials, and he did all he could to overcome the supposed contradictions between the two. He particularly stressed the need for international collaboration and he played a major role in the establishment of the European Ceramics Society, a body that he chaired from 1989 to 1991.

Students and colleagues have praised Hans Hausner's human side, and he enjoyed a high amount of respect as a researcher, which was why numerous national and international honours were bestowed upon him, and also why it was said of him in one conference report that the lecture halls were full when Hans Hausner came to the lectern: "And one never left the room without noting a lasting value."

Lit.: More than one hundred publications in anthologies, conference proceedings and periodicals such as Ceramic forum international.

[F. H.]

Am 13. November 1901 wurde *Heinrich Hertel* in Düsseldorf geboren. Er studierte von 1921 bis 1925, zunächst an der TH München, später an der TH Berlin, Bauingenieurwesen. Anschließend wurde er Assistent, bald Versuchsleiter, an der Deutschen Versuchsanstalt für Luftfahrt in Berlin-Adlershof. Aus der Untersuchung von Flugzeugunfällen entwickelte sich seine Forschung zur dynamischen Belastung und der Ermüdungsfestigkeit von Flugzeugkonstruktionen. 1930 promovierte er an der TH Berlin mit einer Arbeit zur Verdrehsteifigkeit und Verdrehfestigkeit von Flugzeugbauteilen.

1933 wurde Hertel von Ernst Heinkel (1888–1958) engagiert. Im folgenden Jahr wurde er zum technischen Direktor des Heinkel-Flugzeugwerks in Rostock ernannt und übernahm die Leitung der Forschungs- und Entwicklungsarbeiten. Nach einem Streit mit Heinkel wechselte Hertel 1939 zu den Junkers-Flugzeugwerken, wo er Entwicklungschef und Mitglied des Vorstands wurde. Bereits 1938 hatte er an der Universität Rostock eine Honorarprofessur erhalten und 1941 eine Honorarprofessur an der TH Braunschweig. Nach Kriegsende ging Hertel nach Frankreich, wo er an der Entwicklung von strahlgetriebenen Langstreckenflugzeugen und an der Entwicklung der Senkrechtstarttechnik arbeitete.

Nachdem das alliierte Verbot der Luftfahrtforschung in Deutschland aufgehoben worden war, berief die TU Berlin Hertel 1955 auf den neu eingerichteten Lehrstuhl für Luftfahrzeugbau. Zunächst mit nur einer Assistentenstelle ausgerüstet konnte Hertel seine Arbeitsgruppe durch Einwerbung zahlreicher Drittmittel rasch ausbauen. 1960 konnten die drei flugtechnischen Lehrstühle der TU in ein neu errichtetes Institutsgebäude umziehen, für das sich Hertel mit großem Engagement eingesetzt hatte.

An der TU nahm Hertel seine Forschung zur Ermüdungs- und Betriebsfestigkeit von Flugzeugkonstruktionen wieder auf. Seine Ergebnisse fasste er in dem 1969 erschienenen Buch „Ermüdungsfestigkeit der

Heinrich Hertel was born in Dusseldorf on 13 November 1901. He studied civil engineering from 1921 to 1925, first at the TH Munich and later at the TH Berlin. He subsequently became an assistant at the German laboratory for Aviation in Adlershof, Berlin where he was put in charge of testing not long afterwards. Starting with the investigation of aircraft accidents, his research began to extend to the dynamic load and fatigue strength of airplane designs. In 1930 he gained a doctorate from the TH Berlin with a thesis on the rotating rigidity and torsional strength of aircraft construction parts.

In 1933, Hertel was hired by Ernst Heinkel (1888–1958). The following year he was appointed technical director of Heinkel's aircraft plant in Rostock and also became head of research and development there. In 1939, following a disagreement with Heinkel, Hertel moved to the Junkers aircraft factory, where he became head of development and a member of the board. In 1938, he received an honorary professorship at the University of Rostock and in 1941 was also made honorary professor at the TH Braunschweig (Brunswick). When the war was over Hertel went to France, where he worked on the development of long-range jet-powered aircraft and on vertical take-off technology.

Once the allied ban on aviation research in Germany had been lifted, the TU Berlin appointed Hertel to its re-established chair for Aircraft Construction in 1955. Although this only came with one academic assistant post, Hertel was able to rapidly increase his group's size with the aid of generous third-party funding. In 1960 the three chairs of aeronautics at the TU were able to move to a newly established institute building, something that Hertel had helped achieve through his great commitment.

At the TU, Hertel resumed his research into the fatigue and operating strength of airplane constructions. His findings found expression in a book published in 1969, *"Ermüdungsfestigkeit der Konstruktion"* (Resist-

Konstruktion" zusammen, das erste Werk zu diesem Themenkomplex in deutscher Sprache. Seine Arbeiten zur Entwurfs- und Konstruktionstechnik sowie zu neuen Bauweisen und Fertigungsverfahren fanden ihren Niederschlag in dem 1960 veröffentlichten Lehrbuch „Leichtbau", das schnell zum Standardwerk für den Luftfahrzeugbau wurde. Außerdem setzte er seine in Frankreich begonnenen Arbeiten zur Senkrechtstarttechnik an der TU fort.

Richtungsweisend wurde Hertels Forschung auf dem Gebiet der Biologie und Technik. Nach Untersuchung der Körperform verschiedener Fischarten schlug er Anfang der sechziger Jahre eine spindelförmige Rumpfform für die Flugzeugkonstruktion vor. Ein solcher Rumpf hätte bei größerem Nutzraum einen geringeren Luftwiderstand als ein herkömmlicher Zylinderrumpf. Auch wenn Hertels Vorschlag aus technischwirtschaftlichen Gründen bisher nicht in die Praxis umgesetzt worden ist, wurde er mit diesen Arbeiten zum Vordenker der Bionik. Auf diesem Gebiet arbeitete er auch noch nach seiner Emeritierung 1970.

In Würdigung seiner technischen und wissenschaftlichen Leistungen wurde Hertel durch die TU Aachen die Ehrendoktorwürde verliehen, er wurde zum Ehrenmitglied der Deutschen Gesellschaft für Luft- und Raumfahrt sowie zum Ehrenbürger des Raumfahrtzentrums Huntsville ernannt. Hertel verstarb am 5. Dezember 1982 in Berlin.

Lit.: Prof. Dr.-Ing. Heinrich Hertel 65 Jahre. Beiträge aus dem Schülerkreis Professor Hertels, Flugwelt international 18 (1966), S. 872–879, 942–947

[J. Z.]

ance to Fatigue in Design), the first German language work on this issue. His work on the design and production technology as well as on new construction and manufacturing methods were expressed in *"Leichtbau"* (Lightweight Construction) a text book that would quickly become the standard work on aircraft construction (1960). In addition, at the TU Berlin he was to continue his work on vertical take-off technology (which he had begun in France).

Hertel's research in the fields of biology and technology was groundbreaking. After investigating the body shapes of different species of fish, in the early nineteen-sixties he proposed a spindle-like fuselage for airplane construction. Such a fuselage would have a lower air resistance than a conventional cylindrical one and still allow more space for cargo or passengers. Even though Hertel's idea has not yet been implemented for technical and economical reasons, the work made him the intellectual pioneer behind bionics, an area in which he continued working after his retirement in 1970.

In appreciation of his technical and scientific achievements, Hertel was awarded an honorary doctorate by the TH Aachen, and he was also appointed an honorary member of the DGLR (the German Society for Air and Space Travel). He was also made an honorary citizen of the Huntsville space center. Heinrich Hertel died in Berlin on 5 December 1982.

Lit.: Prof. Dr.-Ing. Heinrich Hertel 65 Jahre. Beiträge aus dem Schülerkreis Professor Hertels (Prof. Dr.-Ing. Heinrich Hertel at 65; Contributions from Professor Hertel's circle of students). Flugwelt international 18 (1966), p. 872–879, 942–947

[J. Z.]

Gustav Hertz wurde am 22. Juli 1887 in Hamburg geboren. Nach Besuch des Realgymnasiums studierte Gustav
Hertz ab Ostern 1906 zunächst für zwei Semester in Göttingen und anschließend für ein Semester in München. Nach Ableistung seines Militärdienstes
in Dresden ging Hertz 1908 an die Berliner Universität. Hier schloss er 1911
sein Physikstudium mit einer Dissertation bei Heinrich Rubens (1865–1922)
über das ultraviolette Absorptionsspektrum des Kohlendioxids ab.

Noch 1911 erhielt Hertz eine Assistentenstelle an der Berliner Universität. James Frank (1882–1964), der sich gerade an der Berliner Universität habilitierte, schlug Hertz vor, gemeinsam die Wechselwirkung von
Gasatomen und Elektronen zu untersuchen. Der Frank-Hertz-Versuch lieferte einen unmittelbaren experimentellen Beweis für die diskreten Anre-
gungsniveaus der Elektronen in der Atomhülle. Der Versuch war aber kei-
neswegs zu diesem Zweck konzipiert worden, sondern diente der Untersuchung einer Hypothese über das Stoßverhalten der Elektronen. 1926
erhielten Frank und Hertz für diese Arbeit den Nobelpreis für Physik.

Am ersten Weltkrieg nahm Hertz als Offizier teil. Er wurde schwer
verwundet und arbeitete nach seiner Genesung in der technischen Abteilung für Funkgeräte. 1917 habilitierte er sich an der Berliner Universität mit einer Arbeit „Über den Energieaustausch bei Zusammenstößen
zwischen langsamen Elektronen und Gasmolekülen". Die darin enthalte-
nen Überlegungen zur Bewegung langsamer Elektronen in Gasen können in der Rückschau als Ausgangspunkt einer modernen Plasmaphysik
gewertet werden. 1920 ging Hertz nach Eindhoven (Niederlande), um im
Laboratorium der Philips-Glühlampenfabrik zu arbeiten. Hier setzte er die
Elektronenstoßversuche fort und entwickelte ein Diffusionsverfahren zur
Trennung von Edelgasen.

1926 erhielt er einen Ruf nach Halle, wo er Direktor des Physikalischen Instituts wurde. Ende 1927 nahm Hertz den Ruf an die Techni-

Gustav Hertz was born in Hamburg on 22 July 1887.
Having finished school, he enrolled for two semesters
in Göttingen at Easter 1906 and then for one semester in Munich. He
completed his military service in Dresden and went to Berlin University in
1908. There he finished off his physics studies with a dissertation under
Heinrich Rubens (1865–1922) on the subject of the ultraviolet absorption
spectrum of carbon dioxide.

In 1911, Hertz became an assistant at Berlin University. James Frank
(1882–1964), who had just become a professor there, proposed to Hertz
that the two of them work on the interaction between gas atoms and
electrons. The Frank-Hertz experiment led to immediate experimental
proof of the discrete energy levels of electrons in the atom. The experiment was not designed for this purpose, rather was intended as an investigation of a hypothesis concerning the impulse response of electrons.
Hertz and Frank received the Nobel Prize for Physics for this research in
1926.

Hertz served as an officer in World War I. He was badly wounded
and after his convalescence he worked in the technical division for radios.
He took a professorship in Berlin in 1917 with a paper on the energy
exchange in collisions between slow electrons and gas molecules. The
ideas that he set out in that paper on the movement of slow electrons in
gases can be seen in retrospect as a starting point for the modern field
of plasma physics. Hertz went to Eindhoven in the Netherlands in 1920
to work in the laboratory of the Philips bulb factory. He continued his
experiments in electron collisions there and developed a diffusion process
for the separation of noble gases.

In 1926, Hertz became director of the Physics Institute in Halle.
Then at the end of 1927 he started work at the TH Berlin, where he
was charged with creating a new institute for physics. Within just a few

sche Hochschule Berlin an. Hier war es seine Aufgabe, das neue Institut für Physik aufzubauen. Hertz gelang es in wenigen Jahren, das Fach an der TH gleichrangig zur Berliner Universität zu etablieren. 1931 konnte der Institutsneubau eingeweiht werden. In seiner wissenschaftlichen Arbeit behandelte Hertz die Isotopentrennung durch Diffusion. Es gelang ihm 1932, die Neonisotope 20 und 22 zu trennen sowie schweren Wasserstoff (Deuterium) vom normalen Wasserstoff.

1934 wurde Hertz, da sein Großvater jüdischer Abstammung war, die Lehrbefugnis entzogen. Er verließ daraufhin 1935 die TH und übernahm den Aufbau eines neuen Forschungslaboratoriums bei Siemens. Nach dem Zweiten Weltkrieg arbeitete Hertz in Suchumi (am Schwarzen Meer, UdSSR), er baute dort mit ein physikalisches Institut auf, dessen Aufgabe wesentlich darin bestand, die Grundlage für die Trennung von Uranisotopen im großtechnischen Maßstab zu entwickeln. Von 1954 bis zu seiner Emeritierung 1961 wirkte er dann an der Universität Leipzig. Seit 1954 war Hertz außerdem Mitglied der Akademie der Wissenschaften zu Berlin und von 1963 bis 1968 Sekretar der Klasse Physik und Technik. Er verstarb am 30. Oktober 1975 in Berlin, wohin er nach seiner Emeritierung zurückgekehrt war.

Lit.: Josef Kuczera: Gustav Hertz (Biographien hervorragender Naturwissenschaftler, Techniker und Mediziner, Bd. 80), Leipzig: B.G. Teubner 1985

[J. Z.]

years, Hertz managed to establish an institute on a par with that of Berlin University. A new building for the institute was opened in 1931. In his scientific work in this period, Hertz studied the separation of isotopes by means of diffusion. In 1922 he succeeded in separating neon isotope 20 from isotope 22 as well as heavy hydrogen (deuterium) from normal hydrogen.

In 1934, Nazi race laws meant that Hertz lost his teaching licence because his grandfather was of Jewish descent. He left the TH in 1935 and took on the task of setting up a new research laboratory at Siemens. After World War II, Hertz worked in Sukhumi on the Black Sea coast of the USSR, where he helped establish a physics institute. The foremost task of the institute was to develop a basis for the separation of uranium isotopes in large quantities. From 1954 until he became professor emeritus in 1961, he worked at the University of Leipzig. From 1954 he was a member of the Academy of Sciences in Berlin and he held the honorary position of secretary of the class for physics and technology from 1963 until 1968. He died in Berlin on 30 October 1975, where he had returned to after his years in Leipzig.

Lit.: Josef Kuczera: Gustav Hertz (Biographien hervorragender Naturwissenschaftler, Techniker und Mediziner, Bd. 80), Leipzig: B.G. Teubner 1985

[J. Z.]

Zu des Geistes Flügeln noch die körperlichen wünscht sich nicht nur Faust, vielmehr kündet schon die Antike mit der Sage von Daidalos und Ikaros von dem uralten Traume des Fliegens. Nachdem sich die Menschen 1783 mit einem Heißluftballon erstmals in die Lüfte erhoben hatten, folgten 1891 vor den Toren Berlins Otto Lilienthals Gleitflüge und 1903 der Motorflug der Gebrüder Wright. Das nun zunehmende Interesse am Fliegen und die Einsicht, die Luftfahrt erforschen zu müssen, führten 1912 zur Gründung der Deutschen Versuchsanstalt für Luftfahrt (DVL) in Berlin-Adlershof. Deren Flugzeugabteilung leitete seit dem 1. Januar 1913 *Wilhelm Hoff*, welcher zu dem Zeitpunkt bereits seit einigen Jahren praktisch und wissenschaftlich in der Flugtechnik wirksam gewesen war.

Wilhelm Hoff wurde am 7. Mai 1883 in Straßburg geboren. An der dortigen Universität und an der Technischen Hochschule Berlin studierte er Maschinenbau, arbeitete dann als Ingenieur bei der Motorluftschiff-Studiengesellschaft, danach als Konstrukteur für Luftschiffe und Flugzeuge und war seit 1911 Assistent an der Technischen Hochschule Aachen, wo er 1913 zum Dr.-Ing. promovierte.

Seine Dissertation handelt über *Versuche an Doppeldeckern zur Bestimmung ihrer Eigengeschwindigkeit und Flugwinkel*. Hoff hatte zuvor den Flugzeugführerschein erworben, und nicht nur durch seine Tätigkeit als Flugzeugbauer, sondern auch dank eigener Flüge mit einem Doppeldecker war ihm bewusst geworden, wie wichtig wissenschaftliche Studien seien, um die Konstruktion der Flugzeuge zu verbessern. In Deutschland waren bis Anfang 1911 nur Modellversuche durchgeführt worden, und so kam Hoff der Gedanke, Versuche und Messungen an fliegenden Flugzeugen selbst vorzunehmen; zuvor hatte er die benötigten Messinstrumente eigens zu entwerfen und bauen zu lassen. Hoff betonte den Wert systematischer Stabilitätsprüfungen und beklagte, dass die Stabilitätsvorzü-

It was not only Faust who wanted wings for the body and not only for the soul. On the contrary, in the days of antiquity the saga of Daidalos and Icarus had heralded the age-old dream of flying. After human beings had first conquered the air with a hot-air balloon in 1783, Otto Lilienthal glided for the first time on the outskirts of Berlin in 1891 and the Wright brothers made their first motorized flight in 1903. The increasing interest in flying and the realisation of the need for aviation research led to the foundation of the Deutsche Versuchsanstalt für Luftfahrt (DVL – German Aeronautics Research Institute) in Adlershof, Berlin in 1912. From 1 January 1913, *Wilhelm Hoff* who had been effectively involved in aeronautical engineering for a number of years – both practically and scientifically – ran the organisation's aeroplane department.

Hoff was born in Strasbourg on 7 May 1883. He studied mechanical engineering at the university there and at the Technische Hochschule Berlin, worked first as an engineer at the *Motorluftschiff-Studiengesellschaft* ("Motor Airship Study Society") then as a designer of airships and aeroplanes, and from 1911 onwards was employed as an assistant at the Technische Hochschule Aachen, where he was conferred his doctorate (Dr.-Ing.) in 1913.

His dissertation dealt with "*Versuche an Doppeldeckern zur Bestimmung ihrer Eigengeschwindigkeit und Flugwinkel*" (Trials conducted on biplanes to determine their air speed and air-path azimuth angle). Hoff had earlier acquired a pilot's licence and was well aware of the need for scientific studies to improve the designs of aeroplanes, not just through his activity as an aeroplane constructor but also thanks to experiences as a biplane pilot. Only tests on models had been carried out in Germany up to 1911, so Hoff came up with the idea of performing tests and taking measurements on aeroplanes while in flight – but first he needed to design the necessary measuring instruments himself and then have them made.

ge einer Bauart bisher dem „Empfinden und der Liebhaberei der einzelnen Flugzeugführer" überlassen gewesen seien.

Als Wilhelm Hoff für die DVL tätig war, verfasste er zahlreiche Beiträge zu Fragen verschiedener Gebiete der Luftfahrtforschung und hatte zudem von 1920 bis 1936 die Gesamtleitung der DVL inne, die sich während dieser Zeit als ein Zentrum der deutschen Luftfahrtforschung behauptete; insbesondere gelang es Hoff, dank seiner technischen Übersicht zahlreiche jüngere Kräfte zu gewinnen, die sich als bedeutende Luftfahrtwissenschaftler erweisen sollten. Zum 31. März 1936 schied Hoff aus der DVL aus, um sich nunmehr ganz seiner Professur an der TH Berlin widmen zu können, welche er bereits 1923 angetreten hatte. Die Tätigkeitsberichte der DVL aus den Jahren zuvor melden wiederholt, dass Hoff von seinen Dienstgeschäften an der TH befreit worden sei, weshalb jene Begründung stimmen kann; doch es bleibt offen, ob er diese Entscheidung traf oder treffen musste, weil er nicht mehr wohlgelitten war.

Abschließend seien die Worte eines der engsten Mitarbeiter erwähnt, welcher, als er Wilhelm Hoffs gedachte, neben den fachlichen Leistungen dessen Fähigkeit hervorhob, verschiedenartige Menschen zu sich gegenseitig ergänzender Arbeit zusammenzuführen, und allgemein Hoffs hohe menschliche Eigenschaften samt einer Harmonie aus Leib, Geist und Seele rühmte.

Lit.: Versuche an Doppeldeckern zur Bestimmung ihrer Eigengeschwindigkeit und Flugwinkel. In: Luftfahrt und Wissenschaft 6 (1913). – Zahlreiche Aufsätze, vor allem in der Zeitschrift für Flugtechnik und Motorluftschiffahrt und im Jahrbuch der Deutschen Versuchsanstalt für Luftfahrt.

[F. H.]

Hoff stressed the value of systematic stability testing and complained that the stability benefits of any type of construction had previously been left a matter for the "feelings and fads of the individual pilots".

When Hoff worked for the DVL, he wrote numerous papers on different areas of aviation research and was also responsible for the overall running of the organisation from 1920 to 1936. During this period, the DVL became a centre of German flight research. In particular, thanks to his technical expertise, Hoff succeeded in acquiring the involvement of many younger people who were to later become important aviation scientists. On 31 March 1936, Hoff left the DVL to devote himself fully to his professorship at the Technische Hochschule Berlin, which he had taken up in 1923. The DVL's progress reports testify repeatedly that Hoff had been released from his duties at the TH, which is why this quoted reason may be plausible – however, it remains open as to whether he made this decision himself or was forced to make it because he was no longer well-liked.

Finally, it is worth repeating the words of one of his closest colleagues who, as he recalled Wilhelm Hoff, emphasized not only his technical accomplishments but also his ability to get different types of people involved in complementary tasks, and in general praised Hoff's pronounced human characteristics with their harmony of body, mind and soul.

Lit.: Versuche an Doppeldeckern zur Bestimmung ihrer Eigengeschwindigkeit und Flugwinkel. In: Luftfahrt und Wissenschaft 6 (1913). – Numerous essays, particularly in the Zeitschrift für Flugtechnik und Motorluftschiffahrt and in the Jahrbuch der Deutschen Versuchsanstalt für Luftfahrt.

[F. H.]

Ein Glück sei es für ihn gewesen, dass die TU Berlin ihn 1959 berufen habe, gestand *Walter Höllerer* rückblickend vor einigen Jahren, und noch weitaus mehr war und ist es ein Glück für die TU, für Berlin überhaupt, dass er die Herausforderung angenommen hat und jenem Ruf gefolgt ist.

Walter Höllerer wurde 1922 in Sulzbach-Rosenberg geboren und studierte nach dem Kriege in Erlangen, Göttingen und Heidelberg Theologie, Germanistik, Geschichte, Philosophie und Romanistik. Bekannt wurde er zwar durch seinen 1952 veröffentlichten Gedichtband *Der andere Gast*, doch nicht nur als Schriftsteller erwarb er sich einen Namen, sondern schon bald auch als Lektor, Herausgeber und Literaturwissenschaftler. Und hierbei erwies sich Walter Höllerer als ein solch emsiger und tatkräftiger Anreger, Förderer und Organisator, dass er sich zu einer der wirkungsmächtigsten Personen innerhalb der deutschen Nachkriegsliteratur entwickelte. Allein dass er gemeinsam mit Hans Bender in den Jahren 1954–1967 die *Akzente. Zeitschrift für Dichtung* herausgab, ist wohl hinreichend, um diesen Ruhm zu begründen, doch erwähnt seien noch die seinerzeit richtungsweisende Anthologie *Transit. Lyrikbuch der Jahrhundertmitte* (1956), die 1961 von ihm gegründete Zeitschrift *Sprache im technischen Zeitalter* und das 1963 von ihm ins Leben gerufene *Literarische Colloquium Berlin*.

Von 1959 bis 1987 lehrte Walter Höllerer an der TU Berlin als Professor der Literaturwissenschaft, und auch hier war er über den bloßen Alltagsbetrieb hinaus sogleich nach Amtsantritt als Organisator tätig, indem er die Lesereihe *Literatur im technischen Zeitalter* veranstaltete und über Jahre hinweg deutschsprachige Autoren an die TU Berlin einlud, um sie aus unveröffentlichten Texten vorlesen zu lassen. Die Reihe erhielt einen so überwältigenden Zuspruch, dass die Lesungen mit den nicht deutschsprachigen Autoren seit dem Wintersemester 1961/62 in der Kongresshalle stattfanden.

It was a stroke of luck for him to have been appointed to the Technische Universität Berlin in 1959, admitted *Walter Höllerer* as he reminisced some years ago. It was even more a piece of good fortune for both the TU and Berlin in general that he took up the challenge and followed the call.

Walter Höllerer was born in Sulzbach-Rosenberg in 1922 and studied theology, German studies, history, philosophy and Romance languages and literature in Erlangen, Göttingen and Heidelberg after the First World War. He became well-known thanks to *Der andere Gast* (The other guest) his book of poetry published in 1952. However he not only gained a reputation as a writer, for soon also gained renown as an editor, publisher and literature specialist. Here, Höllerer proved himself to be such an industrious and energetic source of inspiration, patron and organiser that he became one of the most influential people in Post-War German literature. The mere fact alone that he, together with Hans Bender, published *Akzente. Zeitschrift für Dichtung* (*Akzente* – magazine for poetry) from 1954 to 1967 was itself enough to justify this reputation, though the anthology *Transit. Lyrikbuch der Jahrhundertmitte* (Transit. Book of mid-century lyrics) which was trend-setting in 1956, as well as the magazine *Sprache im technischen Zeitalter* (Language in the technical age) that he founded in 1961 and the *Literarische Colloquium Berlin* he set up in 1963 are all worthy of mention.

Walter Höllerer taught at the Technische Universität Berlin as Professor of Literary Studies from 1959 to 1987, and here he was also active immediately in an organisational capacity on top of his day-to-day business, by holding the *Literatur im technischen Zeitalter* (Language in the technical age) series of readings and inviting German-speaking authors to the TU Berlin for years to allow them to read from unpublished texts. The series of events was so well-received so that further readings with non-German-speaking authors were held at the Congress Hall after the 1961/62 winter term.

Eingangs ist von einer Herausforderung die Rede gewesen, und in der Tat sah Walter Höllerer sich damals an der TU Berlin einer innerhalb der deutschen Universitätslandschaft einzigartigen Aufgabe gegenübergestellt. Seine Hörer bestanden nämlich ursprünglich keinesfalls in erster Linie aus angehenden Literaturwissenschaftlern, vielmehr aus künftigen Flugzeugbauern, Chemikern und Elektrotechnikern, weil bis 1968 zu jeder Ausbildung an der TU Berlin unverzichtbar ein humanistisches Studium gehörte. Jene in Vergessenheit geratene Einrichtung ermöglichte es Walter Höllerer, Studenten natur- und ingenieurwissenschaftlicher Fächer eine ergänzende Gegenwelt aus Sprache zu erschließen und ihnen durch Romane und Gedichte neue Bilder zu vermitteln. Dies sollte sich nicht in einem zeitweiligen Prüfungswissen erschöpfen, sondern er setzte auf eine „bleibende Überzeugungskraft der vielstimmigen Literatur". Die Anregungen verliefen nicht einseitig, denn Walter Höllerer bekannte, in den Gesprächen mit diesen Studenten viel mehr Einsichten als in einem traditionellen germanistischen Seminar gewonnen zu haben.

Während nach wie vor das Nebeneinander zweier Kulturen beklagt wird, bleibt dem im Mai 2003 verstorbenen Walter Höllerer das Verdienst, einst an der TU Berlin eine Brücke zwischen Geistes- und Naturwissenschaften gebaut zu haben.

Lit.: Zwischen Klassik und Moderne. Lachen und Weinen in der Dichtung einer Übergangszeit. Stuttgart 1958. – Gedichte 1942–1982. Frankfurt am Main 1982. – Zurufe, Widerspiele. Aufsätze zu Dichtern und Gedichten. Berlin 1993.

[F. H.]

It seemed a major challenge at the start, and at the TU Berlin, Walter Höllerer really did see himself confronted by a task that was then a unique one in the German academic world. To begin with, his listeners did not mainly consist of budding literature specialists, but rather of future aircraft constructors, chemists and electrical engineers because up to 1968, humanistic studies formed an essential part of any course at the TU Berlin. This now forgotten convention made it possible for Höllerer to open up an alternative world of language for science and engineering students and to impart new images to them through novels and poetry. The aim was not to convey the kind of temporary knowledge that sufficed to pass examinations – Höllerer was more concerned with imparting the "lasting power to convince people of the many voices of literature". These stimuli were not just unidirectional, because Höllerer admitted that he had gained many more new insights during discussions with these students than would have been the case in a traditional German studies seminar.

While now, as before, the separation of these two cultures from each other is deplored, Walter Höllerer, who died in May 2003, can take credit for once having built a bridge between the humanities and the sciences at the Technische Universität Berlin.

Lit.: Zwischen Klassik und Moderne. Lachen und Weinen in der Dichtung einer Übergangszeit. Stuttgart 1958. – Gedichte 1942–1982. Frankfurt am Main 1982. – Zurufe, Widerspiele. Aufsätze zu Dichtern und Gedichten. Berlin 1993.

[F. H.]

Eberhard Klitzsch wurde am 18.8.1933 als Sohn eines Revierförsters in Remda/Thüringen geboren. Nach dem Schulbesuch in Weimar, Sondershausen, Schulpforta und Jena schrieb er sich 1951 an der Universität Jena für Geologie ein. Nach 1 1/2 Semestern konnte er nur durch Flucht nach West-Berlin der Verhaftung entgehen, weil er regimekritische Flugblätter verteilt hatte. Er begann nach der Aufnahmeprüfung an der Freien Universität Berlin (FU) das Studium von Neuem; das DDR-Abitur wurde erst aufgrund einer Nachprüfung anerkannt. 1958 wurde er mit einer Arbeit „Das Mitteldevon am Nordwestrand der Dillmulde" an der FU zum Dr. rer. nat. promoviert.

Nach dem Diplom 1957 war er bei einem United States Geological Survey-Team als Ingenieurgeologe tätig. 1958–67 Erdölgeologe bei der DEA-Libyen, war er erst Feldgeologe, dann Explorationsmanager. 1969 habilitierte er sich an der TU Berlin über „Die Strukturgeschichte der Zentralsahara" und wurde 1970 zum Professor ernannt. Hier baute er die angewandten geologischen Disziplinen auf: Hydro-, Erdöl- und Fotogeologie sowie Geologie von Afrika.

Nachdem Klitzsch den Ruf aus Gießen auf eine C4-Professur abgelehnt hatte, war er 1981 maßgeblich an der Gründung des Sonderforschungsbereichs (Sfb) „Geowissenschaftliche Probleme in ariden und semiariden Gebieten" beteiligt und bis 1995 dessen Sprecher. Schwerpunkt seiner Forschungen war die Gliederung der geologischen Großstrukturen, ihrer Wasserressourcen und deren Nutzungsmöglichkeiten. Eingebunden waren etwa 200 Wissenschaftler von TU, FU, Technischer Fachhochschule Berlin und afrikanischen Institutionen. Unter anderem resultierten aus diesem Sfb über 120 Dissertationen, davon 32 afrikanischer Doktoranden. Die Ergebnisse wurden von Klitzsch und Ulf Thorweihe in „Nordost-Afrika: Strukturen und Ressourcen" 1999 herausgegeben. Parallel dazu war Klitzsch federführend an der geologischen Erstaufnahme Gesamtägyptens in 20 Blättern beteiligt. Als Nachfolge des Sfb initiierte

Eberhard Klitzsch was born the son of a forest warden in Remda, Thuringia on 18 August 1933. After attending school in Weimar, Sondershausen, Schulpforta and Jena, he enrolled to study geology at the University of Jena in 1951. After one and a half semesters he had to flee to West Berlin to escape imprisonment for distributing leaflets critical of the East German regime. After taking his entrance examination at the Freie Universität Berlin (the Free University, or "F.U." in West Berlin) he had to start studying again from scratch, as the East German Abitur secondary education certificate was only recognised after he had sat another examination. In 1958 he was conferred a doctorate (Dr. rer. nat.) at the FU with his work Das Mitteldevon am Nordwestrand der Dillmulde (The Middle Devonian Period at the northwestern edge of the Dillmulde).

After receiving his doctorate in 1957 he took part in a United States Geological Survey team as an engineering geologist. From 1958 to 1967, he worked as petroleum geologist for DEA Libya, as a field geologist, then as an exploration manager. He qualified as a university lecturer at the TU Berlin with his work Die Strukturgeschichte der Zentralsahara (The structural history of the Central Sahara) in 1969 and was appointed Professor in 1970. At the TU, he established the applied geological disciplines of hydrogeology, petroleum geology and photogeology, as well as the geology of Africa.

After Klitzsch had turned down the offer of a lucrative "C4" professorship from Gießen, he played a major role in establishing of a special research unit looking into "Geoscientific problems in arid and semi-arid regions" in 1981, remaining its spokesperson until 1995. The main thrust of its research was into the characteristics of the major geological structures, their water resources and their potential uses. The research involved around 200 scientists from Berlin's TU, FU, and Technische Fachhochschule (University of Applied Sciences) as well as from African institutions. For one thing, the special research unit produced no less than 120 dissertations, 32 of which were from African research students. The results were

er den Forschungsschwerpunkt Internationale Geosystemanalyse GEOSYS
und die GeoAgentur Berlin Brandenburg mit dem Ziel, Politik, Wirtschaft
und Wissenschaft für gesellschaftlich relevante Probleme der Geowissenschaften, wie die Veränderungen im Wasserhaushalt, zu sensibilisieren.

Seine intensiven Forschungen in Afrika und Nahost seit 1970 ver-
schafften Klitzsch vielfältige Kontakte zu Institutionen in Entwicklungs-
ländern. Er nutzte sie vor allem für die Aus- und Weiterbildung des wissenschaftlichen Nachwuchses der Regionen: Er initiierte Partnerschaften
der TU mit Hochschulen in Ägypten (Kairo, Assuit, Alexandria) und Somalia (Mogadishu) und 1987 einen Kooperationsvertrag der TU mit der Universität Khartoum. Er versah Kurzzeitdozenturen in Nigeria, Zambia und
im Sudan, in Ife (Nigeria) beteiligte er sich am Aufbau der Angewandten
Geologie; in Khartoum, Lusaka und Ife wurde er zum auswärtigen Prü-
fer ernannt. Als Herausgeber dokumentierte er seine Erfahrungen in „Research in Sudan, Somalia, Egypt and Kenya".

Klitzschs Engagement in der internationalen Wissenschaftsförde-
rung wie in der Forschung wurde 1988 mit dem Bundesverdienstkreuz
und 1999 mit dem Ehrendoktorat der Universität Khartoum gewürdigt;
er ist Ehrenmitglied nicht nur der Deutschen Geologischen Gesellschaft,
sondern auch der Geological Society of Africa.

Lit.: Archiv der TU Berlin

[B. E.]

published by Klitzsch and Ulf Thorweihe in Nordost-Afrika: Strukturen und Ressourcen (Northeast Africa: structures and resources) in 1999. In parallel, Klitzsch also led the first geological survey of all Egypt in 20 maps. He started GEOSYS, a research group concerned with the analysis of international geosystems (as successor to the special research unit) and the Geo-Agentur Berlin Brandenburg with the aim of making the worlds of politics, industry and science more aware of the socially relevant problems of the geosciences, such as the changes in the water balance.

His intensive research work in Africa and the Middle East after 1970 provided Klitzsch with many and diverse contacts to institutions in developing countries. He made use of these contacts above all in providing training and further education for up-and-coming scientists in these regions. He encouraged partnerships between the TU with colleges and universities in Egypt (Cairo, Assuit, and Alexandria) and Somalia (Mogadishu) and in 1987 initiated a cooperation agreement between the TU and the University of Khartoum. He also held short-term lectureships in Nigeria, Zambia and the Sudan. In Ife, Nigeria, he was involved in the establishment of applied geology, while he was also appointed external examiner in Khartoum, Lusaka and Ife. As a publisher, he documented his experiences in "Research in Sudan, Somalia, Egypt and Kenya".

Klitzsch's commitment to promoting science internationally was honoured by the award of the Order of Merit of the Federal Republic in 1988, and in 1999 he was granted an honorary doctorate by the University of Khartoum. He is an honorary member of the Deutsche Geologische Gesellschaft (German Geological Society) and also of the Geological Society of Africa.

Lit.: Archive of the TU Berlin

[B. E.]

Wenn jemandem die Berufswünsche Pfarrer und Naturforscher in die Wiege gelegt werden, um dann nach eigener Aussage eine Art Mittelding zwischen beiden geworden zu sein, handelt es sich um einen philosophischen Mathematiker wie *Adolf Kneser*, der am 19. März 1862 im mecklenburgischen Grüssow als Sohn eines Pfarrers geboren wurde. Zu seinen Paten zählt ein Physiker, der zudem einer der bekanntesten Wissenschaftshistoriker ist, nämlich Johann Christian Poggendorff; und die Sage will, dass dieser sich in seinem Taufspruch sein Patchen als einen Naturforscher wünschte. Doch Kneser studierte Mathematik; zunächst in Rostock, dann in Berlin und Heidelberg, um schließlich wieder in Berlin 1884 bei Leopold Kronecker zu promovieren und sich noch im selben Jahr in Marburg zu habilitieren. Über die Zwischenstation Breslau gelangte er 1889 als Professor der Mathematik ins estnische Dorpat, 1900 wechselte er an die Berliner Bergakademie, die 1916 der TH Berlin angegliedert wurde, und im Jahre 1905 schließlich nach Breslau, wo er bis zu seinem Tode lehrte.

Adolf Kneser hat sich auf mehreren Gebieten der Mathematik ausgezeichnet, und zwar vor allem in der Algebra, der Geometrie, der mathematischen Physik und in der Analysis. Unter dem Einfluss Kroneckers widmete er sich in seiner Dissertation, in seiner Habilitationsschrift und noch in anderen frühen Arbeiten der Algebra, aber schon damals bekunden zwei seiner drei Thesen zur Promotion sein Interesse an der Geometrie. Er verfasste dann auch einige geometrische Schriften über ebene und räumliche Kurven und untersuchte im Rahmen der mathematischen Physik beispielsweise in einigen Aufsätzen die Bewegung eines Massenpunktes in der Umgebung instabiler Gleichgewichtslagen.

Doch besonders hervorgetan hat sich Adolf Kneser in der Analysis; erwähnt seien seine berühmte Untersuchung über die allgemeine Theorie der Sturm-Liouvilleschen Reihen aus dem Jahre 1904 und selbstverständlich sein bekanntes *Lehrbuch der Variationsrechnung*. Es handelte

If someone "inherits" the professions of minister of religion and naturalist as preferred choices of career, and then (as he put it himself) ends up doing something in between, then we can only be talking about the philosophical mathematician *Adolf Kneser*, born the son of a pastor in Grüssow, Mecklenburg on 19 March 1862. His godparents included a physicist who was also one of the most well-known scientific historians, Johann Christian Poggendorff, and the saying goes that, while Kneser was being baptised, Poggendorff expressed the wish that the young Adolf would later become a naturalist. But Kneser studied mathematics – first in Rostock, then in Berlin and Heidelberg, finally obtaining his doctorate under Leopold Kronecker in Berlin in 1884. The same year he qualified as a university lecturer in Marburg. After spending some time in Breslau (now Wroclaw, Poland) he became Professor of Mathematics in Dorpat (today Tartu, Estonia) in 1889. In 1900 he transferred to the Berlin school of mining which was incorporated into the TH Berlin in 1916, before moving back to Breslau in 1905, where he was to teach for the rest of his life.

Kneser excelled in several fields of mathematics, above all in algebra, geometry, mathematical physics, and analysis. Under the influence of Leopold Kronecker, he dedicated himself to algebra in his dissertation, in his postdoctoral thesis and in other early works. Even then, two of the three theses he wrote for his doctorate showed an interest in geometry. He later also wrote several papers on geometry, about plane and spatial curves for example, and (in the context of mathematical physics) investigated the movement of a mass point in the environment of unstable equilibria.

Nevertheless, where Kneser particularly distinguished himself was in analysis. Most worthy of mention is his famous study on the general theory of Sturm and Liouville's series (1904) and, of course, his famous *Lehrbuch der Variationsrechnung* (Textbook of the calculus of variations).

sich um die erste Darstellung des ganzen Gebietes nach langer Zeit, und darum war wohl damals mit aller berechtigten Wertschätzung die Klage über die „spröde Form der Darstellung" verbunden. Doch nach nunmehr über einhundert Jahren lässt sich Knesers Stil als jene herb-strikte Klarheit würdigen, welche als klassische Schönheit mathematischen Werken zu Eigen wird, wenn der Urheber den Stoff überlegen meistert und nur noch gedrängte Substanz in reiner Form ohne jeglichen einleitenden und abschweifenden Zierrat kennt.

Adolf Knesers Wirken erschöpfte sich jedoch nicht in der Mathematik, denn schon während seines Studiums hatte er nebenher Vorlesungen über Geschichte, Literatur und Philosophie gehört und über Erkenntnistheorie nachgedacht. Eine seiner drei Thesen zur Promotion handelt über ein philosophisches Problem, nämlich über allgemeine Naturgesetze und ihr Verhältnis zu den wirklichen Tatbeständen der Natur, und auch später hat er sich wiederholt zu philosophischen Fragen geäußert. Überdies wehte der Geist seines berühmten Paten in ihm, denn Adolf Kneser hat die Wissenschaftsgeschichte stets gewürdigt und die gründliche, auf dem Studium zahlreicher Quellen im Original beruhende Untersuchung *Das Prinzip der kleinsten Wirkung von Leibniz bis zur Gegenwart* beigesteuert, welche Leibniz alle Ehre macht.

Lit.: Mathematik und Natur. Von der Schwere. Zwei akademische Reden. Breslau 1918. – Lehrbuch der Variationsrechnung. Braunschweig 1925 (Erste Auflage Braunschweig 1900). – Das Prinzip der kleinsten Wirkung von Leibniz bis zur Gegenwart. Leipzig und Berlin 1928.

[F. H.]

This was the first depiction of the subject as a whole for a long time, and it was consequently linked to complaints about the "unwieldy form of portrayal", which was then deemed very important. Still, more than one hundred years later, Kneser's style can be praised for a dry and strict clarity that characterises the classical beauty of those mathematical works when the author is patently a master of the subject and only knows how to describe the pure substance of the matter without any introductory or ornamental digression.

Kneser's works were not limited to mathematics though, for even during his studies he also attended lectures on history, literature and philosophy and reflected on knowledge theory. One of this three doctorate theses dealt with a philosophical problem, in fact with general laws of nature and their relationship to the actual realities of nature, and later he also repeatedly discussed philosophical questions. Moreover, the spirit of his famous godfather was still stirring in him, for Kneser had always honoured the history of science and provided a thorough examination based on the study of numerous sources in his original study Das Prinzip der kleinsten Wirkung von Leibniz bis zur Gegenwart (The principle of the smallest effect from Leibniz to the present), which gave great credit to Leibniz.

Lit.: Mathematik und Natur. Von der Schwere. Zwei akademische Reden. Breslau 1918. – Lehrbuch der Variationsrechnung. Braunschweig 1925. – Das Prinzip der kleinsten Wirkung von Leibniz bis zur Gegenwart. Leipzig and Berlin 1928.

[F. H.]

Herbert Kölbel (1908–1995)

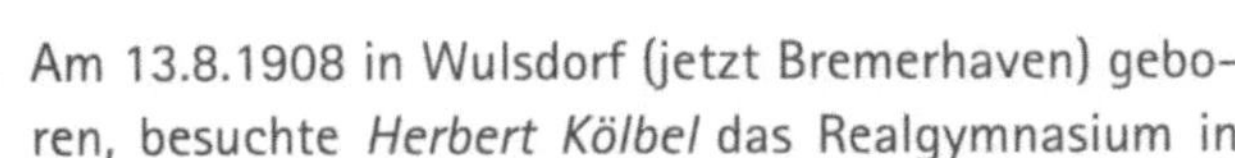

Am 13.8.1908 in Wulsdorf (jetzt Bremerhaven) geboren, besuchte *Herbert Kölbel* das Realgymnasium in Hannover. 1928 begann er das Chemiestudium in Freiburg/Br. bei Hermann Staudinger (1881–1965), 1930 folgte er Walter Hückel (1895–1973) nach Greifswald, bei dem er 1934 promovierte. Als Assistent von Franz Fischer (1877–1947) am Kaiser-Wilhelm-Institut für Kohleforschung in Mülheim/Ruhr lernte er die 1925 entwickelte Fischer-Tropsch-Synthese (FT) kennen, bei der mit aus Kohle gewonnenem CO und H_2 und einem Kobaltkatalysator eine Reihe Kohlenwasserstoffe (KW) als „Kogasin" synthetisiert werden, vor allem für die Benzinsynthese.

1936 ging er zur Rheinpreußen AG Homberg/Rhein als Forschungsleiter für die Inbetriebnahme der ersten großtechnischen FT-Anlage, wurde 1938 Leiter des Synthese-Forschungslabors und ab 1943 Leiter eines Rheinpreußen-Werks. Ihm gelangen Synthesen eines Dieselkraftstoffs aus Kogasin und Teerölen und von Schmierölen für die Kriegsmarine und Reichsbahn. Wegen des Verlusts der Patente, der Zerstörung der Anlagen und Verboten der Besatzungsmacht musste ab 1945 auf Waschmittelproduktion umgestellt werden; für sie entwickelte er bereits abbaubare Tenside. Seine Forschungen über Reaktionsmechanismen im FT-Verfahren führten zu wesentlichen Verbesserungen: In Mehrphasen-Reaktoren ermöglichten suspendierte Katalysatoren eine bessere Wärmeabfuhr und höhere Ausbeuten, im Kölbel-Engelhardt-Verfahren erzielte er 1951 mit dem Einsatz von Eisen statt Kobalt als Katalysator und Wasserdampf statt Wasserstoff eine wesentliche Verbilligung.

1953 wurde Kölbel als Ordinarius an die TU Berlin berufen, wo er die Technische Chemie wiederaufbaute. 1969 konnte der von ihm geplante und benannte Franz-Fischer-Bau bezogen werden. Seine Forschungen an der TU galten weiter der KW-Chemie: Er untersuchte Reaktionsmechanismen der FT-Synthese, den Einfluss der Konstitution von Tensiden auf ihre Eigenschaften und optimale Strukturmerkmale von Schmierölen,

Herbert Kölbel was born on 13 August 1908 in Wulsdorf (now part of Bremerhaven), and attended school in Hanover. In 1928 he started his chemistry studies in Freiburg under Hermann Staudinger (1881–1965), and in 1930 he followed Walter Hückel (1895–1973) to Greifswald, where he took his doctorate in 1934. He worked as an assistant to Franz Fischer (1877–1947) at the Kaiser-Wilhelm-Institut für Kohleforschung (Kaiser Wilhelm Institute for Coal Research) in Mülheim an der Ruhr, where he became acquainted with the Fischer-Tropsch Synthesis. Developed in 1925, it is a process in which carbon monoxide and hydrogen extracted from coal and a cobalt catalyst are synthesised into a series of hydrocarbons such as Kogasin, used especially for gas synthesis.

In 1936 he went to the Rheinpreußen Company in Homberg as a director of research charged with putting into operation the first large-scale Fischer-Tropsch system. In 1938 he became the director of the synthesis research laboratory and in 1943 he became director of one of the Rheinpreußen factories. He succeeded in achieving syntheses of a diesel fuel made of Kogasin and tar oils and of lubricating oils for the navy and railways. After 1945, as a result of losing certain patents, the wartime destruction of the facilities and a prohibition by the occupying powers, the company changed to producing detergents; Kölbel had already developed degradable surface-active agents (tensides). His work on reaction mechanisms in Fischer-Tropsch processes led to significant improvements: in polyphase reactors suspended catalysts enabled improved heat dissipation and higher output; with the Kölbel-Engelhardt process in 1951 he achieved the significant cost-cutting measures of using iron instead of cobalt as a catalyst and steam instead of hydrogen.

In 1953 Kölbel became a professor at the TU Berlin, where he rebuilt the field of technical chemistry. In 1969 the Franz Fischer building, which was planned and named by Kölbel, was opened. His research at the TU

klärte die elektronischen Wechselwirkungen zwischen Gasen und metallischen Katalysatoren und konnte das hydrodynamische Verhalten von Mehrphasenreaktoren mit mathematisch exakten oder halbempirischen Modellgleichungen beschreiben. Daraus entstanden Abschätzungen optimaler Dimensionen und Betriebsbedingungen der Reaktoren, Synthesen höhermolekularen Polymethylens, die einstufige Synthese von Aminen aus Kohlenoxid, Wasser und Ammoniak sowie eine Reihe neuer Schmieröl- und Kunststoffsynthesen. Er gestaltete die Lehre neu, indem er die Wirtschaftschemie zur Prüfung neuer Verfahren, die Theorie der Apparaturen und eine systematische Verfahrenslehre einbezog.

Kölbel engagierte sich als Dekan und Rektor in der universitären Selbstverwaltung und war in wissenschaftlichen und technischen Gesellschaften aktiv. 1967 wurde er in die Akademie der Naturforscher Leopoldina aufgenommen. 1973 emeritiert, starb er am 28.9.1995 in Berlin.

Kölbel bereicherte nicht nur die Technische Chemie um 227 Veröffentlichungen und 433 Patente: Nur eine Verletzung verhinderte seine Teilnahme als Stabhochspringer an den Olympischen Spielen 1936, und mit 180 öffentlichen Konzerten, dem Buch „Über die Flöte" und Noteneditionen unbekannter oder verschollener Werke hat er sich auch in der Musikwelt einen guten Ruf erworben.

Lit.: TU-Archiv

[B. E.]

continued to be concerned with hydrocarbons: he investigated reaction mechanisms of the Fischer-Tropsch Synthesis and the influence of the constitution of tensides on their attributes and the optimal structural characteristics of lubricating oils, he clarified the electron interaction between gases and metallic catalysts and he was able to describe the hydrodynamic behaviour of polyphase reactors with mathematical precision or with semi-empirical comparative models. From these came estimates of the optimal dimensions and operating conditions of reactors, syntheses of higher-molecular polymethylenes, the one-step synthesis of amines from carbon oxide, water and ammonia, and a whole series of new synthetic lubricating oils and other synthetic substances. He overhauled the syllabus to include commercial chemistry for the examination of new processes, the theory of machinery and equipment and a systematic course on processes.

Kölbel was dean and rector at the TU and was active in scientific and technological societies. In 1967 he was inducted into the Akademie der Naturforscher Leopoldina. He became professor emeritus in 1973. He died in Berlin on 28 September 1995.

Kölbel enriched the field of technical chemistry with 227 publications and 433 patents, but he made his mark in other endeavours, too. It was only due to injury that he was unable to participate as a pole-vaulter at the Olympic Games of 1936. He was also prominent in the music world, giving 180 public concerts and publishing a book on the flute "Über die Flöte" and editions of unknown or forgotten musical works.

Lit.: TU-Archive

[B. E.]

Werner Julius March, einer bekannten Fabrikanten- und Architektenfamilie entstammend, wurde am 17.1.1894 in Charlottenburg geboren. Nach dem Abitur am Augusta-Gymnasium in Charlottenburg (1912) studierte er an der TH Dresden Architektur, diente 1914–18 als Soldat und beendete 1918 das Studium an der TH Berlin. Er begann seine Laufbahn als Regierungsbauleiter der Reichsschuldenverwaltung (Berlin-Kreuzberg, Oranienstr.) und war in diesen Jahren Meister-Schüler German Bestelmeyers (1874–1942). Nach dem Regierungsbaumeisterexamen machte er sich 1922 selbstständig. Gemeinsam mit dem Bruder Walter (1900–1969) entwarf er 1926 das Deutsche Sportforum in Berlin, dessen Ausführung Anfang der dreißiger Jahre mit dem Haus des Deutschen Sports beendet wurde. 1928 entwarf er den Umbau des von seinem Vater Otto (1845–1913) 1912 errichteten Grunewaldstadions. Daraus entstand 1932 der Bauauftrag für das zur 11. Olympiade 1936 eröffnete Reichssportfeld mit Olympiastadion – nach Hitlers Einspruch gegenüber dem ursprünglichen Entwurf merklich verändert –, Schwimmstadion und Waldbühne.

Die beeindruckende Anlage brachte weitere Planungsaufträge für Sportanlagen, deren Ausführung durch den Zweiten Weltkrieg verhindert wurde, nur das Stadion in Kairo wurde ab 1956 realisiert. Auch Marchs Entwurf von 1936 für ein archäologisches Museum in Bagdad wurde erst 1952–56 verwirklicht. Am Zweiten Weltkrieg nahm er 1940–45 als Stabsoffizier in der Abwehrabteilung von Admiral Canaris teil.

Nach 1945 wurde March mit dem Wiederaufbau des Mindener Doms und des Mindener Rathauses beauftragt. 1952 entstanden u.a. die evangelische St. Petri-Kirche und 1961 die Vaterunser-Kirche in Berlin-Wilmersdorf (Detmolder Str.), die als besonders gelungener Kirchenbau der Nachkriegszeit gilt. Das Institut für Nachrichtentechnik auf dem Nordgelände der TU entstand 1963–67 nach seinen Plänen. Von den weni-

Werner Julius March was born into a well-known family
of manufacturers and architects, on 17 January 1894
in Charlottenburg, Berlin. He graduated from the Augusta-Gymnasium
secondary school in Charlottenburg in 1912 and went on to study architecture at the TH Dresden. He served as a soldier between 1914 and 1918
and finished his studies at the TH Berlin in 1918. He started his career as
a government clerk for buildings in the district of Kreuzberg, Berlin and
during his years in this role he studied as a master-class student under
German Bestelmeyer (1874–1942). After passing his government master-
grade examination, he established his own business in 1922. With his
brother Walter (1900–1969) he designed the Deutsches Sportforum in
Berlin in 1926, which was completed at the start of the 1930s with the
construction of the Haus des Deutschen Sports. In 1928 he redesigned
Berlin's Grunewald train station, which had been designed by his father
Otto (1845–1913) in 1912. From that came the contract in 1932 to design
the Olympic Stadium for the 1936 Berlin games, the design of which was
significantly altered by Adolf Hitler. The same contract involved designing
a swimming stadium and the Waldbühne open-air concert venue.

These impressive projects brought more work on sports facilities
March's way, but their completion was hindered by World War II. Only his
stadium in Cairo, Egypt was completed in 1956. His 1936 design for an
archaeological museum in the Iraqi capital Baghdad was realised only in
the period 1952–56. During World War II March was a staff officer in the
defence department of Admiral Wilhelm Canaris.

After the war, March was involved in reconstruction projects in the
city of Minden. He designed the Vaterunser church (1961) in Wilmersdorf,
Berlin, which is a particularly well-achieved example of postwar church
design. The Institut für Nachrichtentechnik (Institute for Telecommunica-
tions) in the northern part of the TU Berlin campus was built according
to March's design in 1963–67. He built few residential buildings, but

gen Wohnbauten ist die 1923 fertiggestellte Reichsbanksiedlung in Berlin-Wilmersdorf (Cunostr.) besonders erwähnenswert.

Im Oktober 1953 wurde March zum o. Professor für Städtebau und Siedlungswesen und als einer der Direktoren des Zentralinstituts für Städtebau an die TU berufen. Dieses Institut entwickelte sich zu einem Spezifikum der TU, das durch Vorträge von internationalen Experten zu einem Treffpunkt der Städtebauer, Architekten, Studenten, Politiker und Kollegen anderer Fachrichtungen wurde. Er bemühte sich erfolgreich, das Bundesinstitut für Städtebau nach Berlin zu ziehen. Die TU wurde damit zu einem Zentrum der Stadtplanung. 1960 wurde March emeritiert. Neben einigen ausgeführten Bauten entstanden auch bemerkenswerte Entwürfe, die, wie der des Münchner Olympia-Stadions (1971/72), unausgeführt blieben.

Er hatte wesentlichen Anteil an der Ausarbeitung der Wettbewerbsordnung für Architekten des BDA wie auch der internationalen bei der UNESCO für das Gesamtschaffen der Architekten. Zu bedauern bleibt, dass die von ihm seit 1963 gemeinsam mit der Soziologin Ilse Balg ausgearbeiteten Vorschläge zur Sanierung in Kreuzberg nicht beachtet worden sind und stattdessen der Kahlschlag realisiert wurde.

Die TU ernannte March 1962 zum Ehrensenator. Er starb am 11.1.1976 in Berlin.

Lit.: NDB; Technische Universität Berlin: Akademische Reden, Band 17 (1962); Von Tonwaren zum Olympiastadion: die Berliner Familie March; eine Erfolgsstory. Birgit Jochens und Doris Hünert (Hrsg.), Berlin 2000; TU-Archiv

[M. E.]

the Reichsbanksiedlung in Wilmersdorf, Berlin, which was completed in 1923, is worthy of mention.

In October 1953 March became professor of urban planning and human settlement and also one of the directors of the Zentralinstitut für Städtebau (Central Institute for Urban Planning). This institute later developed into a distinguishing feature of the TU Berlin, as it became a venue for lectures by international experts, including urban planners, architects, students, politicians and others from other fields of expertise. March was successful in his efforts to have the Bundesinstitut für Städtebau (Federal Institute for Urban Planning) located in Berlin. This made the TU a centre for urban planning. March became professor emeritus in 1960. Alongside several completed building designs, March also made several designs that were never realised.

He played an important role in working out rules for architectural competitions for the Bundesvereinigung der Deutschen Arbeitgeberverbände (Confederation of German Employers' Associations). At the international level, he helped UNESCO draw up a similar set of rules. The suggestions put forward by March in collaboration with the sociologist Ilse Balg for the redevelopment of the Kreuzberg district of Berlin were not put into action and the area has suffered as a result.

The TU Berlin made March an honorary senator in 1962. He died i n Berlin on 11 January 1976.

Lit.: NDB; Technische Universität Berlin: Akademische Reden, Band 17 (1962); Von Tonwaren zum Olympiastadion: die Berliner Familie March; eine Erfolgsstory. Birgit Jochens und Doris Hünert (ed.), Berlin 2000; TU-Archive

[M. E.]

Konrad Mellerowicz (1891–1984)

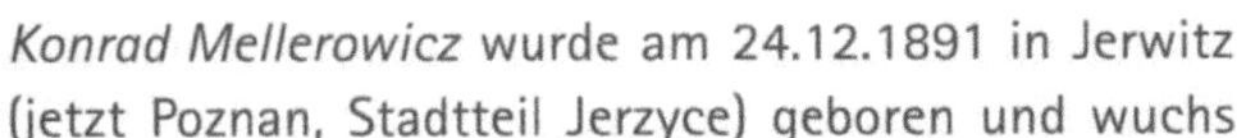

Konrad Mellerowicz wurde am 24.12.1891 in Jerwitz (jetzt Poznan, Stadtteil Jerzyce) geboren und wuchs als Fabrikantensohn in Beuthen/Oberschlesien auf. Im väterlichen Betrieb zum Industriekaufmann ausgebildet, studierte er 1914 Philosophie und Neuphilologie an der Universität Breslau, wurde jedoch bald eingezogen. 1919 ging er er an die Handelshochschule (HHS) Berlin, wo er neben Betriebswirtschaftslehre (BWL) bei Friedrich Leitner (1874–1945) und Willi Prion (s.d.) auch Pägagogik bei Eduard Spranger (1882–1963) und Volkswirtschaftslehre (VWL) bei Werner Sombart (1863–1941) hörte und 1921 das Diplom als Handelslehrer erwarb. 1923 an der HHS Hamburg zum Dr. rer. pol. promoviert, wurde er Assistent von Leitner, bei dem er sich 1926 für das Gesamtgebiet BWL habilitierte. Als Privatdozent vertrat er einen Lehrstuhl der HHS und zeitweise in Kiel, bis er 1929 ein Extraordinariat für Theoretische BWL und BWL des Verkehrs an der HHS Berlin erhielt, an der er 1934 Ordinarius und 1938 Nachfolger Leitners für Allgemeine und Industrielle BWL wurde. An der 1935 in Wirtschaftshochschule (WHS) umbenannten HHS las er über Wertungslehre, Rechnungswesen und Kostentheorie, Industriebetriebslehre, Bankwesen und Betriebslehre des Verkehrs, wobei er stets die Betriebs- in die Volkswirtschaftslehre einband.

Mit der WHS wurde Mellerowicz 1946 an die Humboldt-Universität übernommen. Nach der Äußerung „es gibt keine sozialistische oder liberalistische BWL, hier gibt es nur eine gute oder eine schlechte" entging er 1950 nur durch Flucht der Verhaftung. Schon drei Tage später erhielt er einen Ruf an die TU Berlin, wohin ihm über 160 seiner Ostberliner Schüler folgten, so dass er 1952 eine eigenständige Fakultät für Wirtschaftswissenschaften mit neuen Studiengängen – immer mit Technikbezug – an der TU Berlin begründen konnte. Hier lehrte er neben Theoretischer BWL und Industriebetriebslehre auch Preispolitik, Markenartikelwesen, Werbung, Unternehmenspolitik und Betriebliches Rechnungswesen.

Konrad Mellerowicz was born in Jerwitz (now Poznan, Poland) on Christmas Eve 1891 and brought up in Beuthen (Bytom). He worked on the commercial side of his father's company and in 1914 started studying philosophy and modern linguistics at the University of Breslau (Wroclaw), but he was soon conscripted into military service. In 1919 he went to the Handelshochschule in Berlin, where he studied business under Friedrich Leitner (1874–1945) und Willi Prion (1879–1939, q.v.), pedagogy under Eduard Spranger (1882–1963) and economics under Werner Sombart (1863–1941). He took a diploma as a teacher of commerce in 1921. He earned a doctorate in 1923 from the Handelshochschule in Hamburg, and worked as an assistant to Leitner, with whom he became a professor in the general field of business studies. As a private tutor he worked at the Handelshochschule and temporarily in Kiel. In 1929, he became extraordinary professor for business studies and for business and transportation at the Handelshochschule in Berlin, where he became professor in 1934. In 1938 he succeeded Leitner to take over as professor of general and industrial business studies. He lectured on economic relations, accountancy and cost theory, industrial management, banking and transportation economics, all of which he approached both from a sociopolitical and a business-studies point of view.

The Wirtschaftshochschule was taken over by the Humboldt University in 1946, and Mellerowicz with it. After uttering the opinion in 1950 that "there is no socialist or liberal business studies, here there is only good business studies or bad," he had to go on the run to evade arrest by the communist authorities. Three days later he was summoned to the TU Berlin and more than 160 of his East Berlin students went with him. As a result, he was able to found an independent faculty of economics in 1952 at the TU with new courses, all of them related to technology. Mellerowicz taught theoretical business studies and industrial management, as well as pricing policy, a course on branding, marketing, corporate policy and management accountancy.

Mellerowicz hatte kein enges Spezialgebiet sondern dachte gesamtwirtschaftlich. Kants „es gibt nichts Praktischeres als eine gute Theorie" motivierte ihn zu ständigem Praxisbezug, die reine Mathematisierung hielt er für einen Irrweg, da er für eine sozialverantwortliche Wirtschaftsführung eintrat. Bürgersinn, Verpflichtung gegen die Gemeinschaft, Kontaktfreude und Fairness forderte er nicht nur von seinen Studenten: So lehrte er nach der Emeritierung 1959 als sein eigener Lehrstuhlvertreter weitere 7 Jahre und hielt bis zu seinem Tod durch Herzinfarkt am 25.1.1984 Doktorandenseminare ab. Sein Sohn Harald, Sportmediziner an der FU Berlin, stiftete zum 100. Geburtstag des Vaters den Konrad Mellerowicz-Preis, mit dem alle 2 Jahre hervorragende Arbeiten zur Unternehmensführung ausgezeichnet werden. Ehemalige Schüler ließen zum 111. Geburtstag den maroden Hörsaal 1058 hochmodern in Stand setzen.

Nicht nur für seine über 400 Aufsätze und fast 30 Bücher wurde er 1961 mit dem Großen Verdienstkreuz, dem Goldenen Ehrenring der Deutschen Gesellschaft für Betriebswirtschaft und 1965 der Ehrensenatorwürde der TU Berlin geehrt, sondern auch für sein Engagement in „kampfesfroher Unnachgiebigkeit".

Lit.: Aktuelle Betriebswirtschaft. Festschrift zum 60. Geburtstag von Konrad Mellerowicz, 1952. – Betriebswirtschaftslehre und Wirtschaftspraxis. Festschrift zum 70. Geburtstag von Konrad Mellerowicz, 1961. Konrad Mellerowicz. Bibliographie seiner Veröffentlichungen und Aufsatzsammlung, 1990. – NDB.

[B. E.]

Mellerowicz did not have a narrow field of specialization, rather he thought of economic activity as a totality. Kant's dictum that "there is nothing more practical than a good theory" motivated him always to relate his thinking to economic practice. He regarded pure mathematisation as a mistake because he believed in the social responsibility of those who lead economic affairs. He demanded civic sense, social duty, sociability and fairness from his students. Thus, for the seven years following his becoming professor emeritus in 1959, he continued to work as his own replacement and continued to hold seminars with doctoral students until he died of a heart attack on 25 January 1984. On the 100th anniversary of Mellerowicz's birth, his son Harald, a sports physician at the Free University of Berlin, made an endowment for the Mellerowicz Prize which is awarded every two years to outstanding achievements in the field of business management. On the occasion of the 111th anniversary of his birth, former students had his ramshackle lecture theatre (Room 1058) newly refurbished.

Mellerowicz was awarded the Order of Merit of the Federal Republic of Germany in 1961, the golden ring of honour of the Deutsche Gesellschaft für Betriebswirtschaft and he was made honorary senator of the TU in 1965. These honours were not only given in recognition of his more than 400 academic articles and almost 30 books, but also in recognition of his confrontational and strong-willed personality.

Lit.: Aktuelle Betriebswirtschaft. Festschrift zum 60. Geburtstag von Konrad Mellerowicz, 1952. – Betriebswirtschaftslehre und Wirtschaftspraxis. Festschrift zum 70. Geburtstag von Konrad Mellerowicz, 1961. Konrad Mellerowicz. Bibliographie seiner Veröffentlichungen und Aufsatzsammlung, 1990. – NDB.

[B. E.]

Adolf Miethe wurde am 25.4.1862 in Potsdam geboren, wo sein Vater als Stadtrat wirkte. Schon als Kind interessierten ihn die noch junge Fotografie und optische Instrumente. Zum Studium der Physik, Astronomie und Chemie ging er zunächst nach Berlin, dann nach Göttingen, wo er 1889 mit der Arbeit „Zur Actinometrie photographisch-astronomischer Fixsternaufnahmen" zum Dr. phil. promoviert wurde. Zuvor hatte er bereits am astrophysikalischen Observatorium Potsdam als Hilfsassistent gearbeitet und 1887 (mit einem Freund) das Magnesiumblitzlicht erfunden.

Seit 1889 Wissenschaftlicher Mitarbeiter der Mikroskop-Fabrik E. Hartnack in Potsdam, konstruierte er als einer der ersten mit Jenaer Gläsern Aplanate, die er Anastigmate nannte. Nach Hartnacks Tod in gleicher Position bei Schulze & Bartels in Rathenow, einem Zentrum der optischen Industrie, berechnete und konstruierte er Operngläser, Marine- und Zielfernrohre, wobei ihm, unabhängig von anderen, die eigene Entwicklung eines Teleobjektivs gelang. 1894 ging er zu Voigtländer & Sohn nach Braunschweig. Zunächst mit der Konstruktion terrestrischer Doppel- und Zielfernrohre beschäftigt, wurde er nach der Umwandlung in eine Aktiengesellschaft einer der beiden Direktoren.

1899 wurde Miethe an die TH Berlin berufen, wo sein Vorgänger Hermann Wilhelm Vogel (1834–1898) seit 1873 (Gewerbeakademie) den weltweit ersten Lehrstuhl für Fotochemie, Fotografie und Spektralanalyse inne hatte. Miethe erweiterte das fotochemische Labor, gliederte ihm 1909 eine fotografische Sternwarte an und baute die fotomechanische Reproduktionstechnik aus, u.a. mit einer Versuchsdruckerei. Er initiierte 1921 die Gründung und übernahm die Leitung einer Prüf- und Versuchsanstalt für Kinotechnik; damit führte er dieses Fach an der TH Berlin ein.

Ein Ziel Miethes war die Farbfotografie als Dreifarbensynthese mit drei monochromen Foto-Diapositiven. Nachdem er die rotsensibilisieren-

On 25 April 1862, *Adolf Miethe* was born in Potsdam, where his father was a city councillor. Even as a
child he was interested in the new science of photography and in optical
instruments. He left home to study, first in Berlin, where he studied Physics, Astronomy and Chemistry, before going to Göttingen, where he was
conferred a PhD for his work "*Zur Actinometrie photographisch-astronomischer Fixsternaufnahmen*" (On the actinometry of photographic astro-
nomical images of fixed stars") in 1889. He had previously worked as an
assistant at Potsdam's Astrophysical Observatory and (together with a
friend) invented magnesium flash lighting in 1887.

From 1889 onwards he was employed as a member of scientific
staff at the E. Hartnack microscope factory in Potsdam, where he became
one of the first to design aplanatic lenses using Jena glass, calling them
"Anastigmate". Following Hartnack's death he held a similar post at
Schulze & Bartels in Rathenow, a centre of the optical industry, where
he performed calculations for and designed opera glasses, marine telescopes and telescopic sights, and where he was also able to independently develop a telephoto lens. In 1894 he went to Voigtländer & Sohn in
Braunschweig. Following his initial involvement in the design of terres-
trial binocular telescopes and telescopic sights, he was made one of the
two directors of the firm after it had become a joint-stock company.

In 1899 Miethe was called to the Technische Hochschule in Charlottenburg, Berlin, where his predecessor Hermann Wilhelm Vogel (1834–
1898) had held the first chair for Photochemistry, Photography and Spec-
troanalysis since 1873 (at the *Gewerbeakademie* – Commercial Academy). Miethe expanded the photochemical laboratory there, arranging
for a photographic observatory to be annexed in 1909. He also increased
the amount of photomechanical reproduction technology available there
by adding a test printing department. In 1921 he played a vital role in

den Eigenschaften von Isocyaninen entdeckt hatte, verwendete er „Äthylrot" für panchromatische Trockenplatten und stellte 1909 die Dreifarbenfotografie der Öffentlichkeit vor. Immer an der Verbindung von Fotografie und Astronomie interessiert, studierte er 1908 auf einer Expedition nach Oberägypten Dämmerungserscheinungen und das ultraviolette Ende des Sonnenspektrums, begleitete 1910 die Zeppelin-Expedition nach Spitzbergen und leitete 1914 die Sonnenfinsternisexpedition nach Nord-Norwegen.

Miethes chemische Arbeiten umfassten die synthetische Herstellung von Korunden und Spinellen, die in der Schmuckindustrie Verwendung fanden. 1904 gelang ihm die Synthese von Rubinen für industrielle Zwecke als Lager in Uhren und feinmechanischen Geräten. Ab 1924 beschäftigte er sich mit Versuchen, durch elektrische Entladungen Quecksilber zum Zerfall zu bringen und so Gold zu erzeugen. Die nachgewiesenen geringen Goldmengen erkannten jedoch andere Chemiker als Verunreinigungen.

Miethe gab mehrere Zeitschriften für Fotografie heraus und wandte sich in 13 Büchern und über 90 Aufsätzen sowohl an Fachwissenschaftler wie die interessierte Öffentlichkeit. Künstlerisch begabt, illustrierte er seine Reiseberichte nicht nur mit Fotografien, sondern auch mit eigenen Zeichnungen und Aquarellen. Miethe starb am 5.5.1927 in Berlin an Herzschwäche nach einer unfallbedingten Operation.

Lit.: NDB (dort weitere Literatur)

[B. E.]

founding a test and research institute for cinematography, thus introducing the discipline to the TH Berlin.

One of Miethe's goals was colour photography as a three-colour synthesis using three monochrome photo-diapositives. After discovering the red-sensitising characteristics of the isocyanines, he used " ethyl red" for panchromatic dry plates and was able to introduce trichromatographic photography to the world in 1909. Always interested in linking photography and astronomy, he studied polarisation effects and the ultraviolet end of the solar spectrum on an 1908 expedition to Upper Egypt. In 1910 he accompanied the Zeppelin expedition to Spitzbergen and led the solar eclipse expedition to northern Norway in 1914.

Miethe's chemical activities included the synthetic preparation of corundum and spinels used in the jewellery industry. In 1904, he succeeded in synthesising rubies for industrial purposes for applications such as bearings in clocks, watches and in precision instruments. From 1924 on, he was involved in attempts to make mercury decompose under the effects of electrical discharges in order to create gold. However, other chemists were able to show that the (proven) small amounts of gold present were merely impurities.

Miethe published a number of periodicals on photography, also writing 13 books and more than 90 essays – both for scientific experts and the interested public. Artistically gifted, he illustrated his travel stories not just with photographs, but also with his own drawings and watercolours. Miethe died of cardiac insufficiency in Berlin on 5 May 1927, after an operation following an accident.

Lit.: NDB (additional literature there)

[B. E.]

Am 13. Mai 1851 wurde *Heinrich Franz Bernhard Müller* in Breslau geboren. Nach dem Abitur 1869 und der Teilnahme am Krieg 1870/71 begann er 1871 sein Studium an der Berliner Gewerbeakademie. Nebenbei besuchte er Mathematikvorlesungen bei Elwin Bruno Christoffel (1829–1900) und Karl Weierstraß (1815–1897). Schon als Student bereitete er Kommilitonen von der Bauakademie in Statikrepetitorien auf das zweite Staatsexamen vor.

1875 eröffnete er ein Büro als Zivilingenieur in Berlin. Etwa um diese Zeit erweiterte er seinen Namen durch Anfügen seines Geburtsortes zu Müller-Breslau. Aus seiner Tätigkeit als Repetitor gingen 1875 sein erstes Lehrbuch „Elementares Handbuch der Festigkeitslehre" hervor und zwei Jahre später seine Beiträge über „Elastizität und Festigkeit" sowie „Baumechanik" im Ingenieurtaschenbuch „Hütte". 1883 wurde Müller-Breslau Dozent und 1885 Professor für Bauingenieurwesen an der TH Hannover. Er befasste sich mit der statischen Berechnung von eisernen Brücken, Fachwerken und statisch unbestimmten Tragwerken. Die Ergebnisse seiner Forschung fasste er in seiner zweibändigen „Graphischen Statik der Baukonstruktion" zusammen, deren erster Band 1887 und deren zweiter Band 1891 erschien. Damit wurde Müller-Breslau zum Vollender der klassischen Baustatik.

1888 übernahm Müller-Breslau die Professur für Statik der Baukonstruktionen und des Brückenbaus an der TH Berlin, als Nachfolger von Emil Winkler (1853–1888). Für die Studienjahre 1895/96 und 1910/11 wurde er zum Rektor der TH gewählt. 1901 wurde die „Versuchanstalt für Statik der Baukonstruktion" an der TH eingerichtet, an der Müller-Breslau umfangreiche Untersuchungen über den Erddruck durchführte.

Müller-Breslau blieb Zeit seines Lebens auch als praktischer Bauingenieur aktiv. In Hannover hatte er die Entwürfe für eine Straßenbrücke über die Ihme, eine Markthalle und den Glockenstuhl der Marienkirche

Heinrich Franz Bernhard Müller was born in Wroclaw (Breslau) on 13 May 1851. After completing his *Abitur* examinations in 1869 and taking part in the Franco-Prussian War of 1870/71, he began studying at the Berlin Trade Academy (1871) and attended mathematics lectures given by Elwin Bruno Christoffel (1829–1900) and Karl Weierstraß (1815–1897). Even at that time he was helping fellow students at the Construction Academy prepare for the statics calculations needed for the second state examination.

In 1875 he opened a civil engineer's office in Berlin. Around this time he decided to add the name of his hometown to his surname, becoming known as Müller-Breslau. His activities as student coach helped him in writing his first text book "An Elementary Manual of the Strength of Materials" in 1875, and two years later his contributions on "Elasticity and strength" and on "Structural engineering" appeared in the "Hütte" handbook of engineering. In 1883 Müller-Breslau became a lecturer and in 1885 a professor in civil engineering at the Technische Hochschule in Hanover. He worked on the statics calculations for iron bridges, frameworks, and statically undetermined supporting structures. He summarized the results of this research in his two-volume "Graphical Statics of Building Structure Design", the first volume of which appeared in 1887 and the second in 1891. With this, Müller-Breslau put the finishing touches to classical structural design.

In 1888, he took the chair of Civil Engineering Statics and Bridge Construction at what was then still the Technische Hochschule in Berlin, as successor to Emil Winkler (1853–1888). He was elected principal of the school for the 1895/96 and 1910/11 academic years. In 1901, the Research Institute for the Statics of Structural Design was established at the Technische Hochschule, and it was here that Müller-Breslau conducted extensive investigations into soil pressure.

angefertigt. Außerdem hatte er eine neue Konstruktion für den Bau großer Gasbehälter entwickelt und sich patentieren lassen. In Berlin wurde
ihm 1897/98 die konstruktive Bearbeitung des Berliner Doms vom Fun-
dament bis zu Kuppel übertragen, und er entwarf den Kaisersteg über die
Spree bei Oberschöneweide.

Müller-Breslau befasste sich außerdem seit 1895 mit Luftfahrzeu-
gen. Zunächst beriet er Ferdinand Graf von Zeppelin (1838–1917) bei
der Gestaltung des Tragwerks für dessen Luftschiffe. Während des ersten
Weltkrieges war er Vorsitzender des Fachausschusses für Luftfahrt der
Kaiser-Wilhelm-Stiftung für kriegstechnische Wissenschaften und Berater der Abteilung Luftfahrt des Kriegsministeriums. In dieser Zeit war er
intensiv mit der Berechnung von Tragflächenholmen beschäftigt.

1901 wurde Müller-Breslau zum Mitglied der Preußischen Akademie
der Wissenschaften gewählt, eine besondere Anerkennung, da sich die
Akademie sonst weigerte, Techniker in ihre Reihen aufzunehmen. 1908
wurde er auswärtiges Mitglied der Schwedischen Akademie der Wissenschaften zu Stockholm. Die Ehrendoktorwürde verliehen ihm 1902 die
TH Darmstadt und 1921 die TH Berlin. 1913 wurde er zum Mitglied des
Preußischen Herrenhauses auf Lebenszeit ernannt. Müller-Breslau wurde
1921 emeritiert und verstarb am 23. April 1925 in Berlin-Grunewald.

---

*Lit.: Gebhard Hees: Heinrich Müller-Breslau, Jahrbuch der VDI Gesellschaft Bautechnik
1991, S. 324–370*

[J. Z.]

But Müller-Breslau also remained a working civil engineer all his active life. In Hanover he drew up the draft plans for a road bridge over the river Ihme, a market hall and a bell frame in the *Marienkirche*. In addition, he developed a new design for the construction of large gas-holders, which he was able to patent. In 1897/98 he was handed responsibility for the structural aspects of the design of the Berlin Cathedral – from its foundations to its dome – and he also designed the *Kaistersteg* bridge over the Spree at Oberschöneweide.

From 1895 on, Müller-Breslau also became interested in aircraft. First he advised Count Ferdinand von Zeppelin (1838–1917) on the design and construction of the wing unit for his airships. Then, during the First World War he was chairman of the Expert Committee for Aviation of the Kaiser Wilhelm Foundation for Military Sciences and advisor to the War Ministry's Aviation Section. At this time he was intensively engaged in calculating aircraft wing spar dimensions.

In 1901, Müller-Breslau was elected a member of the Prussian Academy of Sciences, quite a special acknowledgment, since otherwise the Academy refused to accept practical engineers to its ranks. In 1908 he became foreign member of the Swedish Academy of Sciences in Stockholm. He was awarded honorary doctorates by the TH Darmstadt (1902) and the TH Berlin (1921) and in 1913 he was appointed a life member of the upper house of the Prussian parliament. Müller-Breslau was made emeritus professor in 1921 and died on 23 April 1925 in the Grunewald area of Berlin.

Lit.: Gebhard Hees: Heinrich Müller-Breslau, Jahrbuch der VDI Gesellschaft Bautechnik (Annual of the VDI [German Engineers' Association] civil engineering section) 1991, p. 324–370

[J. Z.]

Hans Poelzig (1869–1936)

Poelzig wurde als nichtehelicher Sohn einer thüringischen Gräfin am 30.4.1869 in Berlin geboren, wuchs in einer Kantorenfamilie bei Berlin auf und besuchte das Gymnasium in Potsdam. 1888–93 studierte er Architektur an der TH Berlin. Nach dem Militärdienst trat er 1895 das Referendariat im preußischen Ministerium für öffentliche Arbeiten an und blieb von da an bis zu seinem Tode im Beamtendienst.

1899 legte er die 2. Staatsprüfung ab, wurde 1900 zum Lehrer für architektonisches Zeichnen und Kunsttischlerei an die Breslauer Kunst- und Kunstgewerbeschule berufen und rückte 1903 zum Direktor der Anstalt auf. 1913 zur Akademie erhoben, galt sie in ihrer reformierten Lehrstruktur als wegweisend. 1916–20 arbeitete Poelzig in Dresden als Stadtbaurat und zugleich als Professor an der Akademie.

Mit seiner Berufung an ein Meisteratelier der Akademie der Künste (1920) sowie auf ein Ordinariat für Architektur an der TH (1924) kehrte Poelzig nach Berlin zurück. Er lehrte vor allem Entwerfen von Hochbauten.

Durch Lehrer geprägt, die dem Eklektizismus wie dem Spätklassizismus verbunden waren, entstanden erste Entwürfe, von denen ihm 1896 ein gotisierender Stadthausentwurf den Schinkelpreis einbrachte. Die weiteren Entwürfe sind dann von regionalen Traditionen, vom Jugendstil und von der amerikanischen Geschäftshausarchitektur beeinflusst. Neben einigen Industriebauten (u.a. Chemische Fabrik in Luban bei Posen, 1911/12) entwarf er mit dem Gebäude der Jahrhundertausstellung in Breslau (1913) seinen ersten Repräsentationsbau. In Fachkreisen fiel er jedoch durch den – nicht einmal prämierten – Wettbewerbsentwurf für ein Berliner Opernhaus, den (nicht ausgeführten) Entwurf für das „Haus der Freundschaft" in Istanbul (1916) und das für Max Reinhardt (1873–1943) 1919 am Schiffbauerdamm errichtete Große Schauspielhaus (spä-

Hans Poelzig was born the illegitimate son of a Thuringian countess on 30 April 1869. He was raised near Berlin by the family of a choirmaster and organist and he attended school in Potsdam. He studied architecture at the TH Berlin from 1888 to 1893. After his military service, he took up a training post with the Prussian Ministry for Public Works in 1895, and he remained from that point on in the civil service until his death.

In 1899 he completed his second set of civil service examinations and in 1900 became a teacher of architectonic drawing and cabinet-making at the Kunst- und Kunstgewerbeschule in Breslau (now Wroclaw, Poland). He was made director of the school in 1903. The institution was raised to the status of academy in 1913 and came to be considered ground-breaking for its teaching system. In the period 1916–20 Poelzig worked in Dresden as an advisor on urban planning and simultaneously as a professor at the academy.

Poelzig moved back to Berlin on being awarded the status of master craftsman of the Akademie der Künste (Academy of Arts) in 1920. He became professor of architecture at the TH Berlin in 1924. His teaching there focussed mainly on structural engineering.

His first designs display the influence of teachers who held allegiance to eclecticism and to late classicism; his early design for a gothic-style townhouse won him the Schinkel Prize in 1896. His next designs were influenced by regional traditions, art nouveau and by American office building design. Alongside several industrial buildings (such as the chemical factory in Luban, near the Polish city of Poznan, 1911/12), his building for the exhibition of the century in Breslau of 1913 was his first institutional building. In architectural circles he came to prominence for several designs, including his competition entry for an opera house in Berlin, the never-built Haus der Freundschaft (House of Friendship) in

ter Friedrichstadtpalast, 1986 abgerissen) auf, der viele Ähnlichkeiten mit den Entwürfen für ein neues Salzburger Festspielhaus aufweist.

In Berlin spielte Poelzig in den Kämpfen um die Durchsetzung der Modernen Architektur in den 20er Jahren eine prägende Rolle. Als Mitglied (seit 1908) und zeitweise Vorsitzender des Deutschen Werkbundes (1919–22) sowie als Mitglied der avantgardistischen Architektenvereinigung „Der Ring" hatte er zusätzlich zu seinen Lehrämtern großen Einfluss auf die neue Architektur, wobei sich in den vielen Entwürfen zugleich auch Verharren in manchen spätwilhelminischen Formen und Tendenzen erkennen lässt. Ein für ihn charakteristischer und noch erhaltener Bau ist das „Haus des Rundfunks" (1930) in der Masurenallee. Sein Entwurf für das gegenüberliegende Messegelände (1928/9) blieb unausgeführt. Der Rationalismus seiner Planungen ist sichtbar in dem größten nach seinen Entwürfen entstandenen Baukomplex, dem des Verwaltungsgebäudes der IG-Farben-Industrie in Frankfurt am Main (1929/30). Nicht zuletzt wegen seiner Entwürfe für ein Theater in Charkow und ein Kongressgebäude in Moskau geriet er ins Kreuzfeuer der denunziatorischen Polemiken zu Beginn der 30er Jahre , wurde dem „Kulturbolschewismus" der Avantgarde zugerechnet und sah in Deutschland keine Arbeitsmöglichkeiten mehr. Kurz bevor er in die Türkei emigrieren konnte, starb Poelzig am 14.6.1936 in Berlin.

Sein Sohn Peter (1906–81) wurde ebenfalls Architekt und war 1954–75 Professor für Architektur an der TU Berlin.

Lit.: Theodor Heuss: Hans Poelzig; das Lebensbild eines deutschen Baumeisters, Tübingen 1948; NDB; TU-Archiv

[M. E.]

Istanbul, Turkey (1916), and the Große Schauspielhaus theatre in central Berlin (built 1919, later known as the Friedrichstadtpalast, demolished in 1986). This building, which he designed for Max Reinhardt (1873–1943), displays many similarities to the designs for the Neues Festspielhaus theatre in Salzburg, Austria.

In Berlin Poelzig played a prominent role in the battle over the introduction of modern architecture in the 1920s. As a member (since 1908) and sometime head (1919–22) of the Deutscher Werkbund (German Association of Craftsmen), as a member of the avantgarde architectural society Der Ring, and as the holder of his academic posts, he had an important influence on new architecture. However, the persistence of some late-Wilhelmine forms and tendencies is still recognisable in many of his designs of this period. One extant building that is characteristic of his style is the Haus des Rundfunks (1930) in Masurenallee, Berlin. His design (1928/29) for the convention centre opposite was never realised. The rationalism of his planning is evident in the large administration building complex of the IG Farben company in Frankfurt (1929/30). It was not least for his designs for a theatre in the Soviet city of Kharkov and for a congress hall in Moscow, that he came under fire in the atmosphere of denunciatory polemics of the early 1930s. Accused of the „cultural Bolshevism" of the avantgarde, he saw no professional future for himself in Germany. He died on 14 June 1936 shortly before his planned emigration to Turkey.

Poelzig's son Peter (1906–81) also became an architect and was professor of architecture at the TU Berlin from 1954 to 1975.

Lit.: Theodor Heuss: Hans Poelzig; das Lebensbild eines deutschen Baumeisters, Tübingen 1948; NDB; TU-Archive

[M. E.]

Die Kernspaltung bereitet den Menschen weitaus weniger Schwierigkeiten als die Kernverschmelzung, und wenn sich eine Wissenschaft erst einmal in zwei Teile aufgespalten hat, finden beide nur mühsam wieder zueinander und suchen lieber die Trennung – so die Architektur und das Bauingenieurwesen. Einer gedeihlichen Zusammenarbeit ist das abträglich, weshalb *István Polonyi* durch Worte und Taten versucht, die verloren gegangene Einheit des Bauens weitestmöglich zurückzuerlangen.

István Polónyi wurde am 6. Juli 1930 im ungarischen Gyula geboren; 1952 erhielt er an der TH Budapest das Diplom für Bauingenieurwesen und arbeitete danach an dieser Hochschule und nebenher in der Industrieplanung, bevor er 1956 nach Köln übersiedelte und dort ein Jahr später ein Ingenieurbüro eröffnete. Nachdem er durch einige Arbeiten und insbesondere durch Kirchenbauten mit Dachflächen aus hyperbolischen Paraboloiden bekannt geworden war, berief ihn 1965 die Architekturfakultät der TU Berlin zum Professor der Tragwerkslehre. Anschließend lehrte er von 1973 bis 1995 an der Universität Dortmund als Professor der Tragkonstruktionen.

István Polónyi hat sich schon bei seinen ersten Aufträgen als Ingenieur um eine enge Zusammenarbeit mit dem jeweiligen Architekten bemüht und verweist darauf, dass er sich bei den Kirchenbauten gut in den Architekten hineinversetzen konnte und die Form sogar gemeinsam mit ihm entwickelt hat. Als er 1973 an der Universität Dortmund mit anderen die Fakultät Bauwesen gründen sollte, ergriff er die Gelegenheit beim Schopfe und entwarf mit seinen Kollegen das „Dortmunder Modell Bauwesen": Architekten und Bauingenieure durchlaufen einen gemeinsamen Ausbildungsweg mit jeweils unterschiedlichen Schwerpunkten; und bereits als Studenten lernen sie, mit dem Partner im zukünftigen Berufsleben zusammenzuarbeiten, weil sie sich im Rahmen ihrer Ausbildung mit gemeinsamen Projekten zu beschäftigen haben.

Nuclear fission is much easier to perform than fusion,
and if a science has once split into two parts, they can
only be brought back together with difficulty – so these components prefer to exist separately, like architecture and civil engineering. This is damaging for any successful collaboration, which is why *István Polonyi* tried to recapture the vanished unity of the two building sciences through his words and his deeds.

István Polónyi was born in Gyula, Hungary on 6 July 1930. In 1952 he was awarded a degree in civil engineering at the TH Budapest and he continued working there while also being employed in industrial planning. He emigrated to Cologne in 1956, where he opened a firm of consulting engineers one year later. After he had acquired a reputation on the basis of a number of projects (in particular his church buildings with hyperbolic paraboloid roof areas), the Architecture Faculty of the Technische Universität Berlin invited him to become professor of Structural Frameworks in 1965. He then taught as Professor of Supporting Structures at the University of Dortmund from 1973 to 1995.

Even when busy with his first contracts as an engineer, István Polónyi tried to establish a close cooperation with the architects concerned. He claimed that, with church buildings, he had had no difficulty putting himself in the place of the architect and had even developed the shape together with him. In 1973, when he – with others – was to found the architectural faculty at the University of Dortmund, he grasped the opportunity with both hands, and with his colleagues conceived the "*Dortmunder Modell Bauwesen*" (Dortmund building model): architects and building engineers were to follow a common course of training, though with different points of emphasis. Even as students they were to learn to cooperate with their future trade partners as they had to work on joint projects during their training.

Während all der Jahre als Hochschullehrer hat István Polónyi auch sein Ingenieurbüro betrieben und gezeigt, dass eine Verbindung aus Lehre und Berufspraxis fruchtbar und der Lehre überaus förderlich sein kann. Er konnte so stets aus erster Hand aus dem Alltag der Tragkonstruktionen berichten, über Projekte, Untersuchungen und seine Zusammenarbeit mit den Architekten; insbesondere brauchte er es nicht bei bloßen Beschreibungen der Bauwerkskonstruktionen zu belassen, sondern konnte seinen Studenten darlegen, aus welchen Gründen die Bauwerke so geworden sind.

István Polónyi setzt sich überdies mit grundsätzlichen philosophischen Fragen des Bauingenieurwesens auseinander und hat wiederholt das Verwenden eines unangebrachten Wissenschaftsideals angeprangert. Dem Ingenieur gemäß sei das von Francis Bacon und Galilei begründete induktiv-empirische Vorgehen, das die Baustatik hervorgebracht habe; jedoch nicht das von Pythagoras herstammende deduktive Denkverfahren, wodurch die Baustatik unter dem Einfluss der Leibnizschen Metaphysik in die idealisierte Elastizitätstheorie überführt worden sei. Mit dieser Elastizitätstheorie gehe der Ingenieur nunmehr deduktiv vor, der Architekt jedoch weiterhin induktiv, und darin sieht István Polónyi eine der Ursachen des beklagten Auseinanderdriftens, welches mit dem Dortmunder Modell korrigiert werden könne.

Lit.: Revision des Wissenschaftsverständnisses. Kassel 1986. – Mit zaghafter Konsequenz. Aufsätze und Vorträge zum Tragwerksentwurf 1961–1987. Wiesbaden 1987. – Architektur und Tragwerk (mit Wolfgang Walochnik). Berlin 2003.

[F. H.]

In all his years as a university lecturer, Polónyi continued running his engineering consultancy, showing that combining theory and vocational practice could be productive and extremely beneficial for academia. As a result, he was always able to provide first-hand reports on everyday aspects of supporting structures, on projects and investigations and on his cooperation with architects – in particular, he did not need to restrict himself to the mere depiction of structural design but he could also explain to his students the reasons why buildings had developed the way they had.

Moreover, Polónyi also took a critical look at some fundamental philosophical questions concerning construction engineering and repeatedly denounced the use of a misplaced scientific ideal. According to the engineer, it was the inductive-empirical approach, originated by Francis Bacon und Galileo Galilei, that produced the science of construction statics and not Pythagoras's deductive methodology which transformed statics to the idealised elasticity theory under the influence of Leibniz's metaphysics. With this theory of elasticity, the engineer now proceeds deductively, although the architect continues to progress inductively, and here István Polónyi saw one of the causes of the lamented drifting apart of the two disciplines that could be remedied by means of the Dortmund model.

Lit.: Revision des Wissenschaftsverständnisses. Kassel 1986. – Mit zaghafter Konsequenz. Aufsätze und Vorträge zum Tragwerksentwurf 1961–1987. Wiesbaden 1987. – Architektur und Tragwerk (with Wolfgang Walochnik). Berlin 2003.

[F. H.]

Heinrich Richard Ernst Press wurde am 31. Dezember 1901 in Oker/Harz geboren. In Braunschweig bestand er 1922 sein Abitur und begann ein Bauingenieurstudium an der Technischen Hochschule. Später wechselte er an die TH Berlin, wo er 1926 die Diplomprüfung ablegte. Für kurze Zeit war Press als Mitarbeiter für statische Aufgaben bei Heinrich Müller-Breslau (1851–1925) angestellt. Seine zweijährige Referendarausbildung für die zweite Staatsprüfung zum Regierungsbauführer absolvierte er bei der Reichswasserstraßenverwaltung in Potsdam und der preußischen Wasserwirtschaftsverwaltung in Berlin.

1928 trat Press eine Stelle bei der Gottlieb-Tesch-Bauunternehmung GmbH an. Er stieg hier 1935 zum Prokurist und schließlich zum Direktor und Geschäftsführer auf. 1929 promovierte Press mit einer Dissertation über Fragen der Bodenmechanik an der TH Berlin. Auch während der kommenden Jahre seiner praktischen Ingenieurtätigkeit blieb er wissenschaftlich tätig und verfasste zahlreiche Veröffentlichungen, die sich mit Gründungsproblemen, Betonfertigung und verschiedenen Ingenieurbauten befassten.

1948 erhielt Press den Ruf auf den Lehrstuhl für Wasserbau an der TU Berlin. Mit großem Engagement machte er sich an den Wiederaufbau des schwer beschädigten Laboratoriums für Wasserbau. Bald konnte Press dem Laboratorium durch seine Forschungsarbeit wieder zu Ansehen verhelfen. 1955 wurde das Laboratorium in ein eigenständiges Institut für Wasserbau und Wasserwirtschaft umgewandelt. 1958/59 wurde ein Institutsneubau errichtet, für den Press erhebliche Spendengelder bei Freunden und Gönnern gesammelt hatte. Damit konnte er die wasserbauliche Forschung an der TU auf ein zeitgemäßes Niveau heben.

Auch nach seiner Berufung an die TU hielt Press engen Kontakt zur Praxis. Als Sachverständiger und Gutachter war er an fast allen bun-

Heinrich Richard Ernst Press was born in Oker, in the Harz area of Germany on 31 December 1901. He passed his *Abitur* in 1922 in Braunschweig (Brunswick) and began studying civil engineering at the Technische Hochschule there. Later he transferred to the TH Berlin, where he sat his degree examination in 1926. Press was employed by Heinrich Müller-Breslau (1851–1925) for a short time to help with statics projects. He completed his two-year training for the second state examination at the State Waterways Administration in Potsdam and at the Prussian Water Management Administration in Berlin.

In 1928 Press was given a position at the Gottlieb-Tesch building company and in 1935 he was made the company secretary, finally becoming the managing director. He was awarded a doctorate at the Technische Hochschule Berlin in 1929 with a thesis on questions of soil mechanics. During the years that followed, he remained active in science during his work as an engineer and he wrote numerous publications concerned with foundation problems, concrete manufacturing and other aspects of civil engineering.

In 1948, Press was called to the Chair of Hydraulic Engineering at the TU Berlin. He engaged himself with great commitment in the reconstruction of the badly damaged Hydraulics Engineering laboratory. Soon, through his research work, Press was able to help the laboratory regain its old reputation. In 1955 he converted the laboratory into an independent Institute for Hydraulics Engineering and Water Management. In 1958/59, a new institute building was established, for which Press had collected substantial funds as a result of donations from friends and sponsors. Thus he was able to make hydraulic engineering research at the Technische Universität Berlin state-of-the-art. After his appointment to the TU Berlin, Press remained closely in contact with practical work. As an expert and consultant, he was involved in nearly all the German Feder-

desdeutschen Talsperren- und Wasserkraftprojekten zwischen 1947 und 1967 beteiligt. Auch von ausländischen Regierungen, der Weltbank, der UNO und der FAO wurde er zu den meisten großen Staudammprojekten als Experte herangezogen. Im Laufe seines Berufslebens hat Press so an der Errichtung von mehr als 250 Talsperren mitgewirkt. Außerdem war Press in unzähligen Gremien und Verbänden tätig, so war er unter anderem Präsident des Deutschen Verbandes für Wasserwirtschaft, Herausgeber der Zeitschrift „Die Wasserwirtschaft", Vizepräsident des Internationalen Kongresses für Große Talsperren und ständiger Vertreter der Bundesrepublik Deutschland im Internationalen Schiffahrtskongress.

Trotz seines starken außeruniversitären Engagements hat Press seine Aufgabe als Hochschullehrer sehr ernst genommen. Er verfasste eine Reihe bedeutender Lehrbücher, die häufig in mehreren Auflagen und in verschiedensprachigen Übersetzungen erschienen sind.

Press hat zahlreiche Ehrungen aus dem In- und Ausland erhalten. Ehrendoktoren und Ehrenprofessuren verliehen ihm die Universitäten Braunschweig, Wien, La Paz, Lima, Bologna, Toulouse, Pennsylvania und Thessaloniki. Viele Regierungen zeichneten ihn für die Unterstützung bedeutender Staudammprojekte mit Orden aus, und mehr als 10 Akademien wählten ihn zum Mitglied. Press verstarb am 17. August 1968.

Lit.: Festschrift Heinrich Press, Mitteilungen aus dem Institut für Wasserbau und Wasserwirtschaft Nr. 55, Berlin 1961

[J. Z.]

al Republic's dam and water power projects between 1947 and 1967. He was also consulted as an expert by foreign governments, the World Bank, the UN and the FAO on most of the major dam projects. In this way, Press participated in the construction of more than 250 dams over the course of his working life. Additionally, Press was actively involved in innumerable committees and associations as he was President of the German Federation for Water Management, publisher of the magazine "Water Management", Vice-President of the International Congress for Large Dams, as well as the Permanent Representative of the Federal Republic of Germany at the International Shipping Congress among other things.

Despite his strong commitments outside the university, he took his task as a lecturer very seriously indeed. He wrote a number of important text books that were often frequently republished and translated into a number of languages.

Press received numerous honours from his own country and others. The Universities of Braunschweig, Vienna, La Paz, Lima, Bologna, Toulouse, Pennsylvania and Thessaloniki awarded him honorary doctorates and professorships. Many governments honoured his support for major dam projects with medals, and more than 10 academies elected him to their members. Press died on 17 August 1968.

Lit.: Commemorative Publication on Heinrich Press, Mitteilungen aus dem Institut für Wasserbau und Wasserwirtschaft No. 55, Berlin 1961

[J. Z.]

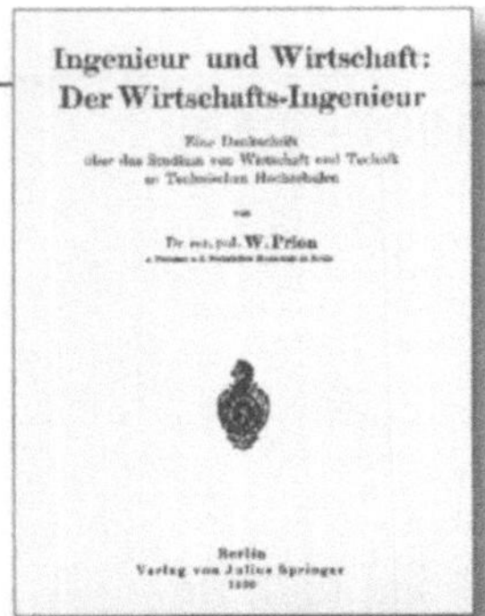

Willi Prion (1879–1939)

Der Kaufmannssohn hugenottischer Herkunft, am 30.11.1879 in Haspe (jetzt Hagen) in Westfalen geboren, verkörperte eine Synthese von französischem Esprit und westfälischer Dickschädeligkeit. Klares Denken, strenge Systematik, geschliffene Rhetorik bis zum ätzenden Sarkasmus, Weltoffenheit, impulsives Draufgängertum vereinte er in sich mit Unbeugsamkeit bis zum Trotz, Zähigkeit, Eigenwilligkeit und konzessionslosem Vertreten seiner oft kühnen Meinungen. Manchmal erschwerte er dadurch das Zusammenleben mit den Kollegen unnötig, doch arbeitete er auch mit ihnen in einer durch Akzeptanz von Kritik und kreative Offenheit geprägten Harmonie zusammen. Der jungen Disziplin Betriebswirtschaftslehre (BWL) vermittelte er einen Teil seines Wesens und beeinflusste zusammen mit seinen „Prioniden" ihre Anfangsrichtung.

Willi Prion besuchte nach einer Banklehre ab 1900 die Handelshochschule (HHS) Leipzig. Neben seiner Berufstätigkeit in der Reichsbank studierte er ab 1902 an der Universität Berlin Staatswissenschaften. 1910 in Freiburg/Br. zum Dr. rer. pol. promoviert, trat er eine Dozentur an der HHS München an, wo er sich für Handelswissenschaften habilitierte. 1913 kam er nach Berlin zurück, als Professor an die HHS. Im Ersten Weltkrieg wurde er als Kompaniechef verwundet und verschüttet. Ab 1917 als Beirat am Reichsschatzamt tätig, wurde er 1920 Ordinarius an der Universität Köln. Dort lehrte er BWL mit Themen ihrer Grenzgebiete zur Volkswirtschaftslehre (VWL): Finanzierung, Bankbetrieb, Börsenwesen, Kreditwirtschaft, Geld- und Währungsprobleme. Als 1925 Oberbürgermeister Konrad Adenauer (1876–1967) jedoch den Prüfungsvorsitz beanspruchte, verließ Prion empört Köln und übernahm ein Ordinariat an der TH Berlin.

Hier erwartete ihn Arbeit auf unbekannten Grenzgebieten. Prion sollte den völlig neuen Studiengang zum Wirtschaftsingenieur einführen, für den er noch 1925/26 einen Studienplan vorlegte, der einer TH

This merchant's son of Huguenot origin, born in Haspe (now Hagen) in Westphalia on 30 November 1879, embodied a synthesis of French *esprit* and Westphalian pigheadedness. He combined clear thinking, strict systematics, and polished rhetoric with an abrasive sarcasm and an openness to what the world had to offer, together with an impulsive daredevilry and tenaciousness that extended up to the point of defiance. He also united a toughness and independence of mind and presented his often bold opinions without making concessions to anyone. Sometimes he made life unnecessarily difficult with his colleagues but he was also able to work with them in harmony, marked by an acceptance of criticism and creative openness. He put a lot of his energy into the new discipline of management studies and together with his "Prionides" was able to influence its direction at the beginning.

Following an apprenticeship at a bank, *Willi Prion* attended the business school in Leipzig from 1900 on. Alongside his work at the Reichsbank he also studied political sciences at the University of Berlin from 1902. In 1910 he was awarded a doctorate (Dr. rer. pol.) in Freiburg/Breisgau. After that, he took up a lectureship at the business school in Munich, where he qualified as a university lecturer for commercial sciences. In 1913 he returned to Berlin as a professor at the business school. He was wounded and trapped during the First World War, in which he served as a company commander. In 1917 he worked as a consultant to the state treasury, and in 1920 became a full professor at the University of Cologne. There he taught management studies and other subjects related to economics: financing, bank operation, the stock exchange, credit and finance, as well as money and currency problems. In 1925 however, Prion left the city in a rage and took a chair at the Technische Hochschule Berlin since the mayor of Cologne, Konrad Adenauer (1876–1967) tried to become chairman of the examination board.

adäquat, den Gegebenheiten der Technik angepasst und pädagogisch
gut durchdacht war. Nach einjährigem Praktikum sollte das 8-semest-
rige Studium die Wirtschaft wissenschaftlich behandeln und zugleich
technisches Denken lehren, damit die Absolventen die Lücke zwischen
Diplomvolkswirten und –kaufleuten einerseits und den Fachjuristen und
–ingenieuren andererseits ausfüllten. Dieses Wirtschaftsstudium, ein Jahr
länger als bisher, stellte erstmals die BWL der bislang dominierenden VWL
gleich. Geschichte und Philosophie der Technik wurden ebenso berück-
sichtigt wie Durchlässigkeit der Studiengänge zwischen TH und Universität. Prion betonte, Wirtschaft könne nicht technisch gehandhabt werden,
die Studenten müssten in zwei unterschiedliche Denk- und Handlungs-
weisen eingeführt werden.

Prion schrieb klar und verständlich, eher große Linien als den täg-
lichen Kleinkram zeichnend, aktuell wie "Kreditpolitik in der Inflation"
1925, systematisch wie die 3-bändige „Lehre vom Wirtschaftsbetrieb"
1935/36, auch popularisierend wie „Börsen – wozu?". Dessen Druckle-
gung erlebte er nicht mehr, Prion erlag am 28.1.1939 überraschend einem
Herzschlag. Während Prions Bücher schnell veralteten, bekam das Wirt-
schaftsingenieur-Studium an der TU Berlin neue Aktualität und wurde
Keimzelle einer neuen Fakultät.

---

*Lit.: W. Hasenack: Zum Gedenken von Willi Prion, in: Betriebswirtschaftliche Forschung
und Praxis 2 (1950), S. 498–504. – NDB.*

[B. E.]

In Charlottenburg he had to deal with tasks in fields that were new to him. It was expected that Prion introduced the completely new industrial engineer course, for which he had already submitted a plan of study for 1925/26. This was suitable for a Technische Hochschule, adapted to the circumstances of the technology and educationally well thought out. After a year's practical work, the course (of eight six-month terms) was to treat economics in a scientific way, at the same time teaching technical ways of thinking, so that it became possible to bridge the gap between political economists and business people on the one hand, and specialist lawyers and engineers on the other. The course, one year longer than was normal at the time, placed business management on the same level as economics, which up to then had been more dominant. History and philosophy of technology were considered, as was the interchangeability of *Technische Hochschule* (polytechnic) and university courses. Prion stressed that economics could not be dealt with by a technical approach, students would have to be introduced to two different ways of acting and thinking.

Prion wrote clearly and understandably, putting his ideas down in broad strokes rather than going into detail about petty day-to-day things. He was up-to-date in regard to such matters as "Credit policies under inflation" in 1925, and also systematic as he showed with the three-volume "Study of economic operation" in 1935/36, as well as with popularizing books like "Stock exchanges – what are they for". Prion did not live long enough to see the latter in print – he surprisingly suffered a fatal heart attack on 28 January 1939. While Prion's books quickly became outdated, the industrial engineer course at the TU gained a new topicality and formed the seedbed of a new faculty.

Lit.: W. Hasenack: Zum Gedenken von Willi Prion, in: Betriebswirtschaftliche Forschung und Praxis 2 (1950), p. 498–504. – NDB.

[B. E.]

Franz Reuleaux (1829–1905)

In Eschweiler bei Aachen kam *Franz Reuleaux* am 30. September 1829 als Sohn eines Mechanikers zur Welt. Nach dem frühen Tod des Vaters 1833 zogen Mutter und Sohn nach Koblenz, wo Reuleaux eine praktische Ausbildung in der Eisengießerei und Maschinenfabrik Zilken erhielt. Ab 1846 arbeitete Reuleaux in der Maschinenfabrik seines verstorbenen Vaters, die ein Onkel übernommen hatte.

Von 1850 bis 1852 studierte Reuleaux Maschinenbau an der Polytechnischen Schule Karlsruhe. Mit seinem Studienkollegen Carl L. Moll (*1831) gab er 1854 die „Konstruktionszeichnungen für den Maschinenbau" heraus. Das Werk erfreute sich bei den Studenten großer Beliebtheit, brachte den Herausgebern allerdings auch einen Plagiatsvorwurf ihres Lehrers Ferdinand Redtenbacher (1809–1869) ein. Von 1852 bis 1854 vertiefte Reuleaux seine naturwissenschaftlichen und philosophischen Kenntnisse durch Studien an den Universitäten in Berlin und Bonn. Anschließend war er in Köln in der Maschinenbauanstalt Baehrens tätig und erhielt 1856 einen Ruf an die Eidgenössische Technische Hochschule Zürich. 1861 erschien die erste Auflage seines Werks „Der Construkteur", mit dem er dem praktisch tätigen Ingenieur ein Werkzeug zur systematischen Konstruktion von Maschinen an die Hand geben wollte.

1864 erhielt Reuleaux einen Ruf an das Königliche Gewerbeinstitut in Berlin. Von 1868 bis 1879 war Reuleaux Direktor des inzwischen zur Gewerbeakademie umbenannten Instituts. 1875 erschien mit dem ersten Band der „Theoretischen Kinematik" Reuleaux' wissenschaftliches Hauptwerk. Reuleaux wollte die Maschinenwissenschaften in eine exakte Wissenschaft verwandeln, die nach deduktiven Methoden vorging. Reuleaux' Kinematik fand sowohl entschiedene Anhänger wie heftige Kritiker.

Auf zahlreichen Weltausstellungen war Reuleaux als Jurymitglied oder als entsandter Reichskommissar tätig. „Billig und schlecht", sein

Franz Reuleaux was born in Eschweiler near Aachen on 30 September 1829. After the early death of his father in 1833, who was a mechanic, he and his mother moved to Koblenz, where Franz received a practical education at the Zilken iron foundry and machine factory. From 1846 he worked in the machine factory of his father, which by then was run by his uncle.

From 1850 to 1852 Reuleaux studied mechanical engineering at the Polytechnic School in Karlsruhe. He and fellow student Carl L. Moll (*1831) published their "Structural Drawings for Mechanical Engineering" in 1854. The work proved very popular among students, but earned the authors an accusation of plagiarism from their teacher Ferdinand Redtenbacher (1809–1869). Between 1852 and 1854 Reuleaux expanded his knowledge of the natural sciences and of philosophy in the universities of Berlin and Bonn. Afterwards he worked at the Baehrens Mechanical Engineering Institute in Cologne and in 1856 he was offered a post at the Eidgenössische Technische Hochschule (Swiss Federal Institute of Technology) in Zurich. In 1861 the first edition of his "Der Construkteur" was published, in which he aimed to provide working engineers with a tool for the systematic construction of machines.

In 1864 Reuleaux was summoned to the Königliches Gewerbeinstitut, a precursor to the TH Berlin. From 1868 to 1879 he was director of the by now renamed Gewerbeakademie. The first volume of his main scientific work, the "Theoretische Kinematik" ("The Kinematics of Machinery"), came out in 1875. Reuleaux wanted to transform knowledge of mechanical engineering into an exact science that could operate according to deductive methods. His approach attracted both strong supporters and determined critics.

Reuleaux served as a jury member at numerous World Exhibitions and as a Royal Commissioner. He caused a sensation when he called the

Urteil über den deutschen Beitrag auf der Weltausstellung in Philadelphia, erregte in Deutschland großes Aufsehen. Er war außerdem Mitglied in der Technischen Deputation für Gewerbe, die die preußische Regierung in technischen Fragen beriet. Er trat in der Deputation mit Entschiedenheit für die Verabschiedung eines Patentgesetzes ein.

In den Jahren 1864 bis 1879 hat sich Reuleaux um die Entwicklung des Otto-Motors verdient gemacht. Er bestärkte Eugen Langen (1833–1895), dem Erfinder Nicolaus August Otto (1832–1891) sein Vertrauen zu schenken, als Mitglied der Technischen Deputation half er bei der patentrechtlichen Absicherung des Erfolges, und er sorgte dafür, dass der Otto-Motor auf der Pariser Weltausstellung 1867 mit der goldenen Medaille ausgezeichnet wurde. Ebenso setzte sich Reuleaux für das Schrägwalzverfahren von Mannesmann zur Produktion von nahtlosen Röhren ein.

Mit dem Zusammenschluss von Gewerbeakademie und Bauakademie zur Technischen Hochschule Berlin wurde Reuleaux Abteilungsleiter für den Maschinenbau und im Studienjahr 1890/91 Rektor der TH. 1896 zog sich Reuleaux ins Privatleben zurück. Die Universität Montreal und die TH Karlsruhe verliehen Reuleaux die Ehrendoktorwürde. Am 20. August 1905 starb Franz Reuleaux in Berlin.

Lit.: Hans-Joachim Braun: Franz Reuleaux, in W. Treue u. W. König: Berlinische Lebensbilder Bd. 6 – Techniker, Berlin: Colloquium Verlag 1990

[J. Z.]

German contribution to the World Fair in Philadelphia "cheap and poor."
He was also a member of the Technical Deputation for Industry, which
advised the Prussian government on technological issues. In this position,
he made a decisive intervention in favour of the adoption of a law on
patents.

Between 1864 and 1879 Reuleaux took part in the development of
the spark-ignition engine. He encouraged Eugen Langen (1833–1895), to
put his confidence in the inventor Nicolaus August Otto (1832–1891). As
a member of the Technical Deputation he helped secure a patent for the
invention and he ensured that the spark-ignition engine was awarded the
golden medal at the Paris World Fair of 1867. Reuleaux also played a role
in developing the Mannesmann rotary piercing process for the production of seamless tubes.

When the Gewerbeakademie and the Bauakademie were amal-
gamated into the TH Berlin, Reuleaux became director of the machine
building division and he became rector in the academic year 1890/91.
He retired from public life in 1896. The University of Montreal and the
TH Karlsruhe awarded him honorary doctorates. Franz Reuleaux died in
Berlin on 20 August 1905.

---

*Lit.: Hans-Joachim Braun: Franz Reuleaux, in W. Treue u. W. König: Berlinische Lebens-
bilder Bd. 6 – Techniker, Berlin: Colloquium Verlag 1990*

[J. Z.]

Alois Riedler kam am 15. Mai 1850 in Graz als Sohn eines Kunstbleichers zur Welt. Er studierte von 1866 bis 1871 Maschinenbau an der TH Graz und wurde anschließend Assistent am Lehrstuhl für Maschinenbau an der Deutschen Technischen Hochschule Brünn. 1873 wechselte er an die TH Wien, wo er zunächst als Assistent und ab 1875 als Maschinenkonstrukteur tätig war. 1880 erhielt Riedler einen Ruf an die TH München, vier Jahre später wechselte er an die TH Aachen, und schließlich wurde er 1888 Professor für Maschinenbau an der TH Berlin.

Riedler schenkte dem Einsatz der Maschinen in der betrieblich-industriellen Praxis immer besondere Aufmerksamkeit. Dies zeigte sich schon in seinen Berichten von den Weltausstellungen in Philadelphia 1876 und Paris 1878, die ihm einige Anerkennung einbrachten. Als einer der Ersten setzte Riedler systematisch Indikatordiagramme zum Vergleich des Wirkungsgrads von Maschinen ein. Bei seiner Berufung an die TH Berlin handelte er aus, dass er an der Hochschule ein privates Konstruktionsbüro einrichten könne. Er entwickelte schnelllaufende Pumpen für den Einsatz in Wasserwerken und zur Wasserhaltung in Bergwerken. Riedlers Expresspumpen waren zwar nach wenigen Jahren durch neuere Konstruktionen verdrängt, die grundlegende Idee, durch höhere Arbeitsgeschwindigkeit zu höherer Effizienz zu gelangen, war jedoch wegweisend für den Maschinenbau.

Auch in der Ingenieursausbildung vertrat Riedler einen konsequent praxisbezogenen Ansatz. Im Anschluss an die Weltausstellung in Chicago 1893 besuchte Riedler zahlreiche technische Lehranstalten in den USA. In einem Bericht empfahl er insbesondere den praxisnahen Unterricht in reich ausgestatteten Ingenieurlaboratorien als vorbildlich und forderte die Einrichtung von Maschinenbaulaboratorien an den Technischen Hochschulen in Deutschland. Bereits 1896 wurde an der TH Berlin das erste Maschinenbaulaboratorium, unter Leitung von Emil Josse

Alois Riedler was born in Graz on 15 May 1850, the son of a bleacher. He studied mechanical engineering at the TH Graz from 1866 to 1871 before becoming an assistant at the Faculty of Mechanical Engineering at the German Technische Hochschule in Brünn. In 1873 he moved to the TH Vienna, where he was active first as an assistant, and then from 1875 onwards as a designer of machines. In 1880 Riedler was called to the TH Munich and four years later moved to the TH Aachen. He was finally appointed Professor for Mechanical Engineering at the Technische Hochschule Berlin in 1888.

Riedler always paid special attention to how machines operated in industrial practice. This had already become apparent in his reports on the World's Fairs at Philadelphia (1876) and Paris (1878) which received some recognition. Riedler was one of the first to systematically use indicator diagrams to compare the efficiency of different machines. When appointed to the TH Berlin, he negotiated that a private engineering design department be set up at the university. He developed high-speed pumps for use in waterworks and in draining mines, though Riedler's express pumps were made obsolete after a few years by later designs. However, the basic idea of achieving higher efficiency by using a higher operating speed was a trailblazing one in mechanical engineering.

Riedler also supported a consistently practical approach toward the training of engineers and he visited numerous technical institutes in the USA following the Chicago World's Fair of 1893. In one report he advocated practical instruction in well equipped engineering laboratories and called for the establishment of mechanical engineering laboratories at German technical universities. The first mechanical engineering laboratory was set up at the TH Berlin as early as 1896, under the direction of Emil Josse (1866–1940). In 1896, Riedler's work "*Das Maschinen-Zeichnen*" (Machine Drawing) reformed the teaching of technical drawing. He propagated accurate, dimensioned black-and-white draughtsman-

(1866–1940), eingerichtet. Den Zeichenunterricht reformierte Riedler mit seinem Werk „Das Maschinen-Zeichnen" (1896). Er propagierte exakte, bemaßte, dem jeweiligen Zweck angepasste Schwarz-Weiß-Zeichnungen und wurde damit zum Begründer des modernen technischen Zeichnens.

Im Jahr 1899 war Riedler Rektor der TH Berlin und führte in dieser Position die Gespräche zur Vorbereitung der Hundertjahrfeier der TH. Gemeinsam mit Adolf Slaby (1849–1913) gelang es ihm, dass Wilhelm II. (1859–1941) den preußischen Technischen Hochschulen im Rahmen dieser Feier das Promotionsrecht verlieh.

Riedler hat sich intensiv mit der Entwicklung des Otto- und des Dieselmotors befasst. 1903 richtete er das „Laboratorium für Verbrennungskraftmaschinen" an der TH Berlin ein, das 1907 zum „Laboratorium für Verbrennungskraftmaschinen und Kraftwagen" erweitert wurde. Riedler übernahm die Leitung dieses Laboratoriums und konstruierte dafür den ersten Rollenprüfstand zur Untersuchung von Kraftwagen.

1897 erhielt er die Grashof-Gedenkmünze, die höchste Auszeichnung des VDI, verliehen. Der Österreichische Ingenieur- und Architektenverein ernannte ihn 1900 zum korrespondierenden Mitglied und verlieh ihm 1931 die Goldene Ehrenmünze. Im Alter von 70 Jahren wurde Riedler 1920 emeritiert. Er kehrte zurück nach Österreich und starb fast vergessen am 25. Oktober 1936.

Lit.: Karl-Heinz Manegold: Alois Riedler, in W. Treue u. W. König: Berlinische Lebensbilder Bd. 6 – Techniker, Berlin: Colloquium Verlag 1990, S. 293–307

[J. Z.]

ship, adapted to the specific purposes in question, and so he became the founder of modern technical drawing.

In 1899 Riedler was appointed principal of the TH Berlin and as such led the discussions on how to prepare for the TH's hundredth anniversary. In this context, Riedler together with Adolf Slaby (1849–1913) succeeded in convincing the Emperor Wilhelm II. (1859–1941) to grant Prussian technical universities the right to award doctorates.

Riedler was actively involved in the development of the Otto (petrol) engine and the diesel engine. In 1903 he set up the Laboratory for Internal Combustion Engines at the TH Berlin, which was expanded to form the Laboratory for Internal Combustion Engines and Motor Vehicles in 1907. Riedler became head of this laboratory and designed the first roller type test stand for it in order to examine motor vehicles.

In 1897 he received the Grashof commemorative medal, the highest honour awarded by the German Engineers' Association, the VDI. The Austrian Engineers' and Architects' Association appointed him honorary member in 1900 and awarded him their Gold Medal in 1931. In 1920, Riedler was elected emeritus professor aged 70. He returned to Austria, to die almost forgotten on 25 October 1936.

Lit.: Karl-Heinz Manegold: Alois Riedler, in W. Treue u. W. König: Berlinische Lebensbilder Bd. 6 (Pictures of Life in Berlin) – Techniker, Berlin: Colloquium Verlag 1990, p. 293-307

[J. Z.]

Hermann Rietschel (1847–1914)

Am 19. April 1847 wurde *Hermann Immanuel Rietschel* als Sohn des berühmten Bildhauers Ernst Rietschel (1804–1861) in Dresden geboren. Nach dem Tod seines Vaters 1861 begann Hermann Rietschel mit 14 Jahren seine Ausbildung in einer Dresdener Schlosserwerkstatt und besuchte ab 1863 die Königlich Polytechnische Schule in Dresden. 1866 trat er in die Maschinenfabrik Eggendorff in Linden bei Hannover ein. Im Sommer 1867 siedelte Rietschel nach Berlin über, wo er sich an der Königlichen Gewerbeakademie immatrikulierte.

Nach Abschluss seines Studiums 1870 arbeitete Rietschel zunächst für die Firma J. & A. Aird. 1871 machte sich er dann mit einem Installationsbetrieb für Heizungs-, Ventilations-, Gas- und Wasseranlagen selbstständig. 1872 gründete er gemeinsam mit seinem zwei Jahre älteren Freund Rudolf Henneberg (1845–1909) die Firma Rietschel & Henneberg. Dank der Kreativität der beiden Gründer entwickelte sich die Firma schnell vom Handwerksbetrieb zum Industrieunternehmen.

Im steigenden Maße wurde Rietschel von Auftraggebern aus der öffentlichen Verwaltung als Berater in Fragen der Gesundheitstechnik herangezogen. Im Januar 1880 beteiligte sich Rietschel an der Gründung des Verbands deutscher Ingenieure für Heiz- und Gesundheitstechnische-Anlagen. Im gleichen Jahr trennte er sich von seinem Freund und Geschäftspartner und begann eine zweite Laufbahn als Zivilingenieur und Wissenschaftler. Neben zahlreichen Entwurfsarbeiten und Gutachten begann Rietschel, auch publizistisch tätig zu werden. Die für 1882 geplante Hygiene-Ausstellung in Berlin wurde wesentlich von Rietschel vorbereitet. Nachdem die gesamte Ausstellung einen Tag vor der geplanten Eröffnung nieder brannte, schaffte es Rietschel, sie innerhalb eines Jahres zu rekonstruieren. Ende 1883 wurde Rietschel in Anerkennung seiner wissenschaftlichen Verdienste der Professorentitel verliehen.

Hermann Immanuel Rietschel was born in Dresden on 19 April 1847, the son of the famous sculptor Ernst Rietschel (1804–1861). After his father died, Hermann Rietschel embarked at the age of 14 on his training in a metal workshop. He started at the Royal Polytechnic School in Dresden in 1863. In 1866 he began work at the Eggendorff machine factory in Linden near Hanover. In the summer of 1867 he moved to Berlin where he entered the Königliche Gewerbeakademie, a precursor to the TH Berlin.

Having finished his studies in 1870, Rietschel first worked for the J.&A. Aird Company. In 1871 he started his own business installing heating, ventilation, gas and plumbing systems. In 1872 he and his friend Rudolf Henneberg (1845–1909) founded the company Rietschel & Henneberg. Thanks to the creativity of the two founders, the company was soon transformed from a handicraft business into an industrial concern.

Rietschel was increasingly approached by clients from the public sector seeking his advice in matters of public health engineering. In January 1880 he became a co-founder of the Confederation of German Engineers for Heating and Sanitary Systems. In the same year he ceased working with his partner and friend and embarked on a second career as a civil engineer and a scientist. Alongside numerous design jobs and other research, Rietschel also became involved in public relations. The hygiene exhibition planned for 1882 in Berlin was prepared largely by him. After the entire exhibition was burned to the ground one day before its official opening, Rietschel managed to rebuild it within a year. In recognition of his services to science, he was awarded the title of professor at the end of 1883.

On 13 July 1885 Rietschel took the world's first chair in ventilation and heating systems at the TH Berlin. In 1887 a testing centre for heating and ventilation systems was annexed to his professorship and numerous

Am 13. Juli 1885 wurde Rietschel auf den weltweit ersten Lehrstuhl für Ventilations- und Heizungswesen an der Königlichen Technischen Hochschule zu Berlin berufen. 1887 wurde dem Lehrstuhl eine Prüfstation für Heizungs- und Lüftungsanlagen angegliedert, zahlreiche Forschungsthemen wurden hier bearbeitet – so unter anderem zur Bestimmung von Wärmekoeffizienten und Wärmeabgabe von Dampf- und Wasserheizkörpern, Versuche über Luftströmungen in Rohren und Kanälen sowie Eichung von Messinstrumenten der Heiz- und Raumlufttechnik. 1907 konnte ein Neubau für das Labor bezogen werden. Rietschels gesammelte Erfahrung floss in den ‚Leitfaden zum Berechnen und Entwerfen von Heizungs- und Lüftungsanlagen' ein, der 1893 erschien.

In den Studienjahren 1889/90 und 1899/1900 wurde Rietschel zum Dekan der Abteilung Architektur gewählt und 1893/94 zum Rektor der TH Berlin. Rietschels Leistung wurde vielfach gewürdigt, so unter anderem durch einen Ehrendoktor der Technischen Hochschule Dresden, die Wahl zum korrespondierenden Mitglied der Kgl. Schwedischen Akademie der Wissenschaften sowie zum Ehrenmitglied des österreichischen Ingenieur- und Architektenvereins und des Royal Sanitary Institute in London. 1910 wurde Rietschel emeritiert, am 18. Februar 1914 verstarb er in Berlin nach langer schwerer Krankheit.

Lit.: Klaus W. Usemann: Entwicklung von Heizungs- und Lüftungstechnik zur Wissenschaft: Hermann Rietschel – Leben und Werk. München; Wien: Oldenbourg 1993

[J. Z.]

research projects were carried out there, including measuring the thermal coefficient and heat loss of steam- and water-radiators, tests on air flows in pipes and ducts, and calibrating measuring instruments related to heating and ventilation. A new building was constructed to house the laboratory in 1907. Rietschel drew on all his experience for his "Guide to Calculating and Designing Heating and Ventilation Systems," which was published in 1893.

In the academic years 1889/90 and 1899/1900 Rietschel was chosen as dean of the architecture department and in 1893/94 he was made rector of the TH Berlin. He received many official acknowledgements for his achievements, including an honorary doctorate from the TH Dresden, a nomination as a corresponding member of the Royal Swedish Academy, and honorary membership of the Austrian Society of Engineers and Architects and of the Royal Sanitary Institute in London. Rietschel was made professor emeritus in 1910, and he died on 18 February 1914 in Berlin after a long illness.

Lit.: Klaus W. Usemann: Entwicklung von Heizungs- und Lüftungstechnik zur Wissenschaft: Hermann Rietschel – Leben und Werk. Munich; Vienna: Oldenbourg 1993

[J. Z.]

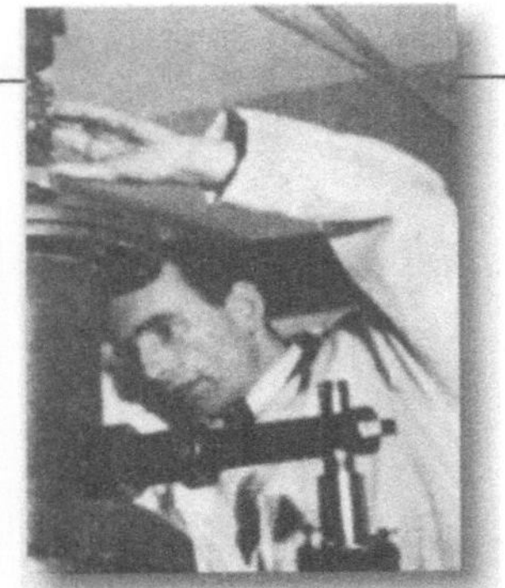

Ernst Ruska (1906–1988)

Ernst Ruska wurde am 25. Dezember 1906 in Heidelberg geboren. Nach Abschluss des Gymnasiums begann er 1925 an der TH München das Studium der Elektrotechnik und wechselte 1927 an die TH Berlin. 1928 wurde er Mitarbeiter in der Arbeitsgruppe von Max Knoll (1897–1969) am Hochspannungsinstitut, die sich mit der Konstruktion eines Kathodenstrahloszillografen befasste. Hier sollte er die Fokussierung des Elektronenstrahls untersuchen und entwickelte in diesem Kontext eine elektronenoptische Anordnung zur vergrößerten Abbildung von Objekten. Ermutigt durch diesen Erfolg konstruierte Ruska mit Unterstützung von Knoll 1931 ein Elektronenmikroskop aus zwei magnetischen Linsen, das allerdings nur eine 17-fache Vergrößerung erreichte. 1933 konnte Ruska dann mit einer verbesserten Version eine Auflösung erzielen, die diejenige eines Lichtmikroskops übertraf. Die Eigenschaften magnetischer Elektronenlinsen kurzer Brennweite waren dann auch Thema seiner Dissertation von 1934.

Ende 1933 trat Ruska eine Stelle bei der Fernseh AG in Berlin an, wo er Bildempfangs- und Bildsenderöhren entwickelte. Gemeinsam mit Bodo von Borries (1905–1956), der ebenfalls in Knolls Arbeitsgruppe promoviert hatte, suchte Ruska nach industriellen Geldgebern, um aus der Versuchsapparatur ein serienreifes Elektronenmikroskop zu entwickeln. Schließlich überzeugten sie 1936 die Siemens & Halske AG und konnten ein Laboratorium für Elektronenmikroskopie bei Siemens einrichten. Ende 1939 wurde das erste Elektronenmikroskop ausgeliefert, und bis Kriegsende folgten etwa 35 weitere. 1944 habilitierte sich Ruska an der TH Berlin.

Nach dem Krieg musste Ruska zunächst die durch Demontage aufgelöste Entwicklungs- und Produktionsabteilung bei Siemens neu aufbauen. 1949 konnten wieder Elektronenmikroskope geliefert werden und 1954 wurde das Elmiskop I vorgestellt, das erstmalig eine Auflösung unter 10 Å erreichte und mit dem Siemens seine Führungsrolle auf dem Markt

Ernst Ruska was born in Heidelberg on 25 December 1906. After his schooling at the Gymnasium he started studying electrical engineering at the TH Munich in 1925, before moving to the TH Berlin in 1927. In 1928 he began work on the construction of a cathode-ray oscilloscope as a member of Max Knoll's (1897–1969) working group at the Institute for High-Voltage Engineering. Here he examined how the electron beam focussed and in this context he developed an electron-optical arrangement for the enlarged display of objects. Encouraged by this success, in 1931 Ruska (with Knoll's support) designed an electron microscope comprising two magnetic lenses, though this could only achieve an enlargement of 17x. In 1933, with an improved design, Ruska was then able to obtain a resolution exceeding that achieved with an optical microscope. The characteristics of magnetic electronic lenses of short focal length formed the topic of his thesis in 1934.

In late 1933, Ruska joined the "Fernseh AG" television company in Berlin, where he was to develop tubes for receiving and transmitting images. Together with Bodo von Borries (1905–1956), who had also gained a doctorate in Knoll's group, Ruska looked for industrial sponsors to help develop an electron microscope for series production, based on the available test equipment. Finally, in 1936 they were able to win over Siemens & Halske AG as backers and set up a laboratory for electron microscopy at the Siemens plant. The first electron microscope was produced towards the end of 1939, and around 35 more were to follow before the war had ended. In 1944 Ruska qualified as a university lecturer at the TH Berlin.

After the war Ruska first had to re-establish the development and production department, which had been dissolved following the disbanding of Siemens. In 1949 the production of electron microscopes recommenced and in 1954 the ELMISKOP I was introduced – a resolution of less than 10 Å was possible for the first time, enabling Siemens to regain

zurückerobern konnte – bis 1965 wurden über 1000 Instrumente verkauft. Daneben engagierte sich Ruska verstärkt in der wissenschaftlichen Forschung. 1947/48 leitete er eine Abteilung für Elektronenmikroskopie am Akademieinstitut für Medizin und Biologie in Berlin-Buch. 1948 übernahm er die Leitung einer Arbeitsgruppe am Fritz-Haber-Institut (FHI) in Dahlem. Seit 1949 lehrte er als Privatdozent an der TU Berlin und als Honorarprofessor an der FU Berlin. Ebenfalls 1949 wurde er auf der Gründungsversammlung der Deutschen Gesellschaft für Elektronenmikroskopie in Düsseldorf zu deren erstem Vorsitzenden gewählt.

1955 gab Ruska seine Industrietätigkeit bei Siemens auf. Er übernahm den Aufbau und die Leitung einer unabhängigen Forschungsabteilung für Elektronenmikroskopie am FHI, aus der zwei Jahre später das Institut für Elektronenmikroskopie wurde. Hier entwickelte er unter anderem das Einfeldkondensorobjektiv, mit dem später die meisten hochauflösenden Elektronenmikroskope ausgerüstet wurden. 1959 wurde er zum außerplanmäßigen Professor an der TU Berlin ernannt.

Ruska wurden in Laufe seines Lebens vier Ehrendoktortitel verliehen, verschiedene Akademien und wissenschaftliche Gesellschaften haben ihn zum Ehrenmitglied gewählt und er erhielt zahlreiche Preise, darunter zuletzt als höchste Auszeichnung 1986 den Nobelpreis für Physik. Ruska verstarb am 27. Mai 1988 in Berlin.

Lit.: L. Lambert und T. Mulvey: Ernst Ruska (1906–1988), Designer Extraordinaire of the Electron Microscope: A Memoir, in Advances in Imaging and Electron Physics 95 (1996), S. 1–62

[J. Z.]

its role as market-leader. Over 1000 of these instruments had been sold by 1965. At the same time, Ruska was becoming increasingly committed to scientific research. In 1947 and 1948 he headed a department for electron microscopy at the Academy Institute for Medicine and Biology in Berlin's Buch district. In 1948 he took charge of a working group at the Fritz-Haber-Institut (FHI) in Dahlem. From 1949 onwards, he taught as an untenured assistant professor at the TU Berlin and as an honorary professor at the Free University of Berlin. In 1949 he was also elected the first chairman of the German Society for Electron Microscopy at its founding meeting in Dusseldorf.

In 1955, Ruska relinquished his industrial activities with Siemens. He took over responsibility for establishing and running an independent research department for electron microscopy at the FHI, out of which the Institute for Electron Microscopy was born two years later. Here, among other things, he developed a single-field condenser lens, a device with which most high resolution electron microscopes were later to be fitted. In 1959 he was appointed extraordinary professor at the TU Berlin.

In his lifetime, Ruska was awarded four honorary doctorates, a number of academies and scientific associations made him their honorary member and he received other numerous prizes, culminating in the highest honour possible – the 1986 Nobel Prize for physics. Ruska died in Berlin on 27 May 1988.

Lit.: L Lambert and T. Mulvey: Ernst Ruska (1906–1988), Designer extraordinaire of the Electron Microscope: A memoir, in "Advances in Imaging and Electron Physics" 95 (1996), p. 1–62

[J. Z.]

„Schaff Schiffe, oder vielmehr: Dem Himmelswind gemäße Segel richte her", rief Kepler dem Entdecker der Jupitermonde begeistert zu: Menschen werde es geben, die schräken selbst vor jener Ödnis nicht zurück. Er sah, dass nicht so sehr ein besonderes Fahrzeug zu entwerfen und zu bauen sei, sondern dass vor allem zu überlegen sei, wie man sich außerhalb irdischer Gefilde fortbewegen könne. Zudem galt es, erst einmal die Schwere zu überwinden, welche die Träumenden einstweilen noch mit beiden Beinen auf dem Boden der Erde stehen ließ. Der zündende Funke entsprang am Ende des 19. Jahrhunderts aus dem Antriebsprinzip der Feuerwerksraketen und entfachte Anfang der 1930er Jahre die ersten genauen mathematischen Untersuchungen über ein Raumflugzeug, das Geschwindigkeiten von Mach 10 in einer Höhe von 60 bis 70 km erreichen sollte.

Deren Urheber *Eugen Sänger* wurde am 22. September 1905 im böhmischen Preßnitz geboren. Während seiner Jugend hatte ihn Kurd Laßwitz' Science-Fiction-Klassiker *Auf zwei Planeten* beeindruckt, und obschon er 1923 an der TH Graz den Weg zum Bauingenieur eingeschlagen hatte, belegte er bald nach einem Wechsel an die TH Wien zusätzlich das Fach Flugwissenschaft, und zwar nachdem er Hermann Oberths Buch *Die Rakete zu den Planetenräumen* gelesen hatte. Sänger promovierte 1930 an der TH Wien, blieb dort als Assistent und veröffentlichte 1933 seine bisherigen Forschungsergebnisse in seinem Buch *Raketenflugtechnik*, welches seinem Anspruch, das Problem des Raketenfluges in ernst zu nehmende Bahnen zu lenken, in beeindruckender Weise gerecht wird. Eugen Sänger hob darin in einer Aufzählung einiger „unmittelbar praktischer Aufgaben" des Raketenfluges zudem hervor, dass dieser „notfalls eine Kriegswaffe von außerordentlicher Wirkung bilden" solle.

Oberth hatte die Möglichkeit eines Fluges außerhalb der Erdatmosphäre samt den zu bewältigenden Herausforderungen aufgezeigt. Das

"Build ships or rather: set your sails to the wind"
Kepler, excitedly cried out to Galileo Galilei, the dis-
coverer of the moons of Jupiter. "There will be people who will not fear
even that wretchedness". He saw that it was not so important to design
and build a particular vehicle but that it was above all worth considering
how man could be conveyed beyond the earthly skies. Moreover, it was
first necessary to overcome gravity, something that left the dreamers
with both feet on the planet earth. The spark of ignition came at the end
of the 19th century with the principle used to propel firework rockets, and
this was kindled in the early nineteen-thirties with the first exact mathematical investigations into a space vehicle designed to travel at speeds
of Mach 10 and a height of 60 to 70 km.

The pioneer behind all this was *Eugen Sänger*, who had been born
in the Bohemian city of Preßnitz (now Prísecnice, Czech Republic) on
22 September 1905. In his youth, he had been astounded by a classic
science fiction story by Kurd Laßwitz called *Auf zwei Planeten* (On two
planets). Although he had chosen to become a civil engineer on beginning his studies at the Technische Hochschule in Graz in 1923, soon after
going to the TH in Vienna he also enrolled for aviation science – after
reading Hermann Oberth's *Die Rakete zu den Planetenräumen* (The rocket
to planetary space). Sänger was awarded a doctorate at the Technische
Hochschule in Vienna in 1930, and remained there as an assistant before
publishing his previous research findings in his book *Raketenflugtechnik*
(Rocket aviation) in 1933, which did spectacular justice to his claim to
be able to reduce the problem of rocket flight to a practicable size. In a
list of immediate practical assignments for rocket flight, Eugen Sänger
emphasized that, if necessary, it could be employed as a "weapon of war
with an extraordinary effect".

Oberth outlined the possibility of flight beyond the earth's atmosphere, including all the challenges that had to be met. The main prob-

Hauptproblem bereitete der Antrieb; es hieß, die Brennstoffkammern der Rakete zu optimieren und einen geeigneten Kraftstoff zu finden, dessen Heizwert vor allem eine möglichst hohe Auspuffgeschwindigkeit ergeben sollte, und hier setzte Sänger mit seinen Untersuchungen an. Nachdem er sich 1936 entschieden hatte, jene im Deutschen Reich fortzuführen, erhielt er ein Jahr später den Auftrag, in Trauen ein Institut zur Erforschung der Raketentechnik einzurichten, wo er überdies begann, das Staustrahltriebwerk zu entwickeln. Seit 1942 setzte er seine Forschungen in Ainring fort, nach dem Kriege erst in Frankreich und seit 1954 in Deutschland. Neben seinen einfallsreichen Weiterentwicklungen des Antriebes widmete er sich seinem Lieblingsvorhaben eines wiederverwendbaren Raumflugzeuges, welches nur bis zu einer Außenstation im Weltraum fliegen sollte. 1963 berief die TU Berlin Eugen Sänger auf den Lehrstuhl für Raumfahrt; er starb jedoch bereits 1964.

Sänger wurde nicht müde, Grundlagenfragen der Raumfahrt zu betrachten und dabei in die Zukunft zu blicken. So erörterte er schon 1956 das Prinzip einer relativistischen Antriebsmechanik, des so genannten Photonenstrahlantriebes; der Treibstoff wird in Strahlungsquanten verwandelt, die mit Lichtgeschwindigkeit ausgestrahlt werden. Einstweilen ein Traum, doch der nüchternen Träumer bedarf die Weltraumfahrt.

Lit.: Raketenflugtechnik. München und Berlin 1933. – Zur Mechanik der Photonen-Strahlantriebe. München 1956. – Raumfahrt: heute – morgen – übermorgen. Düsseldorf und Wien 1963.

[F. H.]

lem was caused by the method of propulsion. The goal was to optimise the rocket's fuel chambers and find a suitable source of energy with a calorific value that provided as high an exhaust gas speed as possible – this is where Sänger began his investigations. After deciding in 1936 to continue his research within the German Reich, he received the call one year later to establish an Institute for Rocket Technology Research in Trauen, where he also began to develop the ram jet engine. After 1942 he continued his researches in Ainring, then (after the war) in France and from 1954 on in Germany. Alongside his imaginative further development of this form of jet propulsion, he also devoted his energies to a favourite project – a re-useable space vehicle that would fly to a distant external space station. In 1963, the Technische Universität Berlin invited Eugen Sänger to occupy the Chair of Astronautics. However, he died shortly afterwards in 1964.

Sänger never tired of confronting fundamental questions of space travel and so always looked to the future. For instance, as early as 1956 he discussed the principle of a relativistic propulsion mechanism, the so-called photon drive, where the fuel would be converted into radiation quanta, emitted at the speed of light. This still remains a dream for the time being, but then space travel does call for serious dreamers.

Lit.: Raketenflugtechnik (Rocket aviation) Munich and Berlin, 1933. – Zur Mechanik der Photonen-Strahlantriebe. (On the mechanics of the photon drive) Munich 1956. – Raumfahrt: heute – morgen – übermorgen (Space travel: today, tomorrow and the day after) Dusseldorf and Vienna, 1963.

[F. H.]

Hans Scharoun (1893–1972)

Als ein Arkadien im Herzen Berlins tadelte ein Kritiker das Kulturforum und sprach damit ein wahres Wort. Dort hat nämlich *Hans Scharoun* seine schon Anfang der Zwanzigerjahre erhobene Forderung verwirklicht, dass in der Gegenwart wieder Stätten der Gemeinschaft und Stadtwahrzeichen zu schaffen seien, wohin die Sinne der Bewohner gerichtet seien. Die Krone aber sollte ein der Kunst gewidmetes Haus sein, und in der Tat erweist sich die Philharmonie als sein Meisterwerk.

Hans Scharoun wurde am 20. September 1893 in Bremen geboren, wuchs in Bremerhaven auf und studierte von 1912 bis 1914 Architektur an der TH Berlin. Nach dem Ersten Weltkrieg arbeitete er bis 1925 als freier Architekt im ostpreußischen Insterburg, danach lehrte er in Breslau und kehrte schließlich 1932 als freier Architekt nach Berlin zurück, wo er zudem von 1946 bis 1958 an der TU Berlin den Lehrstuhl für Städtebau innehatte. Während dieser Jahre an der TU Berlin veranstaltete er Kolloquien zum Städtebau, die regen Zuspruch fanden, und er legte Wert darauf, dass sein Institut ein Lehr- und ein Forschungsinstitut sein sollte, um den Studenten zu ermöglichen, sowohl an praktischen als auch an Forschungsaufgaben mitzuarbeiten.

Inmitten der Aufbruchsstimmung des Neuen Bauens während der Zwanzigerjahre hatte Hans Scharoun seinen Stil zu entdecken begonnen und unter dem Einfluss Hugo Härings zum organischen Bauen gefunden. Wie dieser betont er, dass eine „tragende Idee" vorhanden sein müsse, um den Entwurf eines Bauwerkes zu bestimmen; diese Idee erschöpft sich nach Scharoun nicht in einer bloßen Zweckerfüllung, sondern im Gegensatz zu einem äußerlich-starren, mechanischen Prinzip liegt gleichsam eine sich selbst entwickelnde, innere Gesetzlichkeit eines lebendigen Baues vor.

One critic castigated the Kulturforum, an ensemble of galleries, museums and concert halls in the western part of Berlin, as an arcadia in the heart of Berlin, and in doing so he spoke the truth. In fact it was there that Hans Scharoun made his call (first expressed in the early nineteen-twenties) for contemporary sites for the community and city landmarks to be created, to guide the consciousness of the citizens. Pride of place was to be given to a building dedicated to the arts, and in truth the resulting Philharmonie was to become his masterpiece.

Hans Scharoun was born in Bremen on 20 September 1893, grew up in nearby Bremerhaven and studied architecture at the Technische Hochschule Berlin between 1912 and 1914. Following World War I, he worked as a freelance architect in Insterburg, East Prussia (now Chernachovsk, Russian Federation) until 1925, before taking up teaching at Breslau (now Wroclaw, Poland) and finally returning to Berlin as a freelance architect in 1932. In Berlin he was also to hold the Chair of Urban Development at the Technische Universität from 1946 to 1958. During this period he organised colloquia on urban development at the Technische Universität, which found a great resonance, and he considered it important that his institute be both a teaching and research institution, allowing students to take part in practical as well as research tasks.

In the midst of the dawning Neues Bauen (New Building) movement during the nineteen-twenties, Scharoun had begun to discover his own style, and under the influence of Hugo Häring found his way to the school of organic building. Like Häring, he stressed that a "supportive idea" had to be present in order to determine a building's design. According to Scharoun, this idea did not just consist of a mere fulfilment of a purpose – on the contrary, it comprised an externally rigid, mechanical principle. There had to be, so to speak, a living building, developing within itself and with an internal legitimacy.

Berühmte Beispiele dieser Architektur zeigt das erwähnte Kulturforum. Menschen schließen sich zu einem Kreis zusammen, wenn unerwartet Musik erklingt; eine wohl bescheidene Beobachtung, die einen Hans Scharoun jedoch zur Idee einer Philharmonie inspiriert, wie sie einfacher und genialer nicht sein könnte: Er verlegt jenen Vorgang in einen Konzertsaal und lässt die Musik auch räumlich und optisch das Herz bilden. Um dieses Podium der Musiker sind die Zuhörer konzentrisch angeordnet, und dank einer nie gehörten Akustik strömen von einem Jungbrunnen neubelebende Klänge hinauf.

Gegenüber der Philharmonie liegt der erst nach dem Tode Hans Scharouns vollendete Neubau der Staatsbibliothek. Ein größeres und zu Mauerzeiten zweites Gebäude war erforderlich, Hans Scharoun aber entwarf voll visionären Wirklichkeitssinnes eine Bibliothek für ein zukünftiges Berlin ohne Mauer. Eine Behörde übernahm jedoch noch zu seinen Lebzeiten die Oberaufsicht und halbierte die Leserzahl, welche dem ursprünglichen Plan zu Grunde gelegen hatte. Trotz weiterer schwerer Eingriffe hat doch zumindest die Idee einer Leserbibliothek überdauert: Kein beklemmender Gerichtsaufgang voll einschüchternder Stufen bedrückt, keine düster-schweren Türen schotten gegeneinander enge Gebiete wie dumpfe Stücke zersprungenen Gutes schroff ab – sanft gleiten Leser in den weiten Saal, in den einen weiten Saal, welcher wieder in den Einklang weitet, sanfter gleitet ein Leser weiter in die lichte Stille: Der Geist ist frei, frei wie Fittiche des Himmels.

Lit.: Struktur in Raum und Zeit. In: Handbuch moderner Architektur. Berlin 1957. – Vom Stadt-Wesen und Architekt-Sein. Berlin 1986. – Bauten, Entwürfe, Texte. Hrsg. von Peter Pfankuch. Berlin 1993.

[F. H.]

Hans Scharoun (1893–1972)

The Kulturforum is a well-known example of this architecture. When music is heard unexpectedly, people join up to form a circle. Even though this is a particularly unpretentious observation, it was the principle that inspired Scharoun to design a Philharmonie that could not have been simpler or more ingenious. He transposed this process into a concert hall and let the music form its heart – both spatially and optically. The audience is arranged concentrically around the podium where the musicians stand, and revitalized sounds emanate upwards as if from a fountain of youth, thanks to acoustic features never encountered before.

Opposite the Philharmonie is the new Staatsbibliothek (state library) building, only completed after Scharoun's death. While the Berlin Wall still stood, a larger second building was needed. But, always the visionary, Hans Scharoun designed a library for a future Berlin without a Wall. However, during his lifetime, an authority gained overall supervision and halved the amount of space available to readers planned originally. Despite further major intrusions, the idea of a reader-oriented library at least still survived: there were no hampering courthouse-style staircases full of intimidating steps and no dismal heavy doors demarcating tightly enclosed areas. Instead, readers glide through the vast reading room, and another broad hall (itself expanding into harmony) lets visitors softly glide into the bright peace and quiet. The spirit is free, like the wings of heaven.

Lit.: Struktur in Raum und Zeit. (Structure in space and time) In: Handbuch moderner Architektur (Handbook of modern architecture) Berlin 1957. – Vom Stadt-Wesen und Architekt-Sein (Of the nature of the city and of being an architect). Berlin 1986. – Bauten, Entwürfe, Texte. (Buildings, designs and texts) published by Peter Pfankuch. Berlin 1993.

[F. H.]

Georg Schlesinger, der am 17. Januar 1874 in Berlin geboren wurde, entstammte einer jüdischen Familie. Nach seinem Abitur absolvierte er ein einjähriges Praktikum in einer Mechanikerwerkstatt, bevor er sich 1892 an der Technischen Hochschule Berlin immatrikulierte. Sein Studium des Maschinenbauwesens schloss er 1897 mit einer prämierten Diplomarbeit über Dampfmaschinensteuerung ab.

Bereits seit März 1897 arbeitete Schlesinger bei der Werkzeugmaschinenfabrik Ludwig Loewe & Co AG als Konstrukteur. 1902 übernahm er bei Loewe die Leitung der Konstruktionsabteilung. Besondere Aufmerksamkeit schenkte Schlesinger der Werkstatt-Messtechnik, er schuf das erste funktionsfähige Grenzlehren- und Passungssystem.

Über das Thema Passungssysteme für den Werkzeugmaschinenbau promovierte Schlesinger 1904 an der TH Berlin. Noch im selben Jahr wurde er auf den neu gegründeten Lehrstuhl für Werkzeugmaschinen, Fabrikanlagen und Fabrikbetriebe an der TH berufen. Ab 1906 baute er hier das Versuchsfeld für Werkzeugmaschinen auf, das beispielgebend für vergleichbare Lehrstühle in Deutschland wurde. Schlesingers Forschungstätigkeit war breit angelegt, von der Konstruktion von Werkzeugmaschinen, über die Einbindung der menschlichen Arbeitskraft in den Produktionsprozess, bis zur Fabrikorganisation. Zahlreiche Unternehmen beriet Schlesinger bei der Konzipierung neuer Anlagen und der Umstrukturierung der Produktion.

Während des 1. Weltkrieges leitete Schlesinger die Königliche Gewehrfabrik in Spandau und war wesentlich an der Gründung einer Prüfstelle für Ersatzglieder beteiligt. Hier wurden Prothesen unter dem Aspekt der beruflichen Wiedereingliederung von Kriegsbeschädigten getestet und neue Prothesen entwickelt. Nach dem Ersten Weltkrieg wurden an Schlesingers Lehrstuhl Pionierarbeiten auf den Gebieten der

Georg Schlesinger was born to a Jewish family in Berlin on 17 January 1874. After finishing school, he did a one-year apprenticeship in a mechanics workshop before enrolling in the TH Berlin in 1892. He qualified in mechanical engineering in 1897 with a prize-winning dissertation on control gears for steam engines.

Schlesinger had already started working as a design engineer at the Ludwig Loewe & Co AG machine tool factory in March 1897. In 1902 he became head of the Design Engineering Department at Loewe. Schlesinger paid particular attention to the measuring technology in the workshop, and he created the first functioning limit gauge and fit classification system. He took his doctorate at the TH Berlin in 1904 for his work on a fit classification system for machine tool engineering. In the same year, he was promoted to the new professorial chair of machine tools, manufacturing systems and plant operation at the TH Berlin. From 1906 he established experimental practices in the field of machine tools that became a standard for similar institutes across Germany. Schlesinger's research interests were wide, ranging from design engineering to machine tools and to the integration of human labour into the production process and to factory organisation. He advised many companies on the organisation of new facilities and on restructuring existing production systems.

During World War I Schlesinger was director of the Royal Gun Factory in Spandau and he was central to the foundation of a test centre for artificial body parts. The centre developed new prostheses suited to the occupational rehabilitation of wounded soldiers. After World War I, Schlesinger's department was behind pioneering work in the areas of scientific plant management and rationalised production techniques. In 1924 he travelled in the USA to study developments in machine tool engineering and in the car industry. He highlighted in numerous lectures and publications the possibilities that American production processes offered. These ideas found expression in his work as an advisor to industry. In 1918

wissenschaftlichen Betriebsführung und einer rationalisierten Produktionsweise geleistet. 1924 reiste Schlesinger durch die USA, um die Entwicklungen im Werkzeugmaschinenbau und der Automobilindustrie zu studieren. In zahlreichen Vorträgen und Veröffentlichungen zeigte Schlesinger Möglichkeiten auf, von den amerikanischen Produktionsverhältnissen zu lernen. Auch in seine Tätigkeit als Industrieberater floss diese Erfahrung ein. 1918 gründete Schlesinger außerdem eine Arbeitsgruppe für industrielle Psychotechnik, aus der 1922 ein eigenständiges Institut hervorgehen sollte. Ziel war es, die Erkenntnisse der Experimentalpsychologie für die Optimierung des industriellen Arbeitsprozesses nutzbar zu machen.

Nach Machtantritt der Nationalsozialisten 1933 musste Schlesinger wegen seiner jüdischen Abstammung seinen Lehrstuhl an der TH aufgeben. Außerdem wurde er im April 1933 unter dem haltlosen Vorwurf der Spionage verhaftet und saß sieben Monate in Untersuchungshaft. Im März 1934 emigrierte er in die Schweiz, wo er als Gastdozent an der ETH Zürich wirkte. Im November 1934 siedelte er nach Brüssel über, wo er als Industrieberater und Dozent an der Université Libre arbeitete. 1939 emigrierte Schlesinger nach Großbritannien. Hier richtete er ein Forschungslaboratorium für Fertigungstechnik in Loughborough ein, das er bis 1944 leitete. Noch am Morgen seines Todestages, am 6. Oktober 1949, schloss er das Manuskript für sein letztes Werk, die „Messung der Oberflächengüte", ab.

Lit.: Günter Spur, Wolfram Fischer (Hrsg.): Georg Schlesinger und die Wissenschaft vom Fabrikbetrieb. München; Wien: Carl Hanser 2000

[J. Z.]

Georg Schlesinger (1874–1949)

Schlesinger founded a working group on industrial psychology, out of which grew an independent institute in 1922. The goal of the discipline was to harness the knowledge gained through experimental psychology in order to optimize the industrial process.

When the Nazis came to power in 1933, Schlesinger had to give up his professorship because of his Jewish background. In April of that year, unsubstantiated accusations of spying were levelled against him and he was arrested and detained for seven months. In March 1934 he emigrated to Switzerland, where he worked as a visiting lecturer at the Eidgenössische Technische Hochschule (Swiss Federal Institute of Technology) in Zurich. In November 1934 he moved to Brussels, where he worked as an industrial consultant and as a teacher at the Université Libre. In 1939 Schlesinger emigrated to Great Britain. There he set up a research laboratory devoted to manufacturing engineering in Loughborough, where he remained director until 1944. On the morning of the day he died, 6 October 1949, he completed the manuscript for his final work, "The Measurement of Surface Quality".

Source: Günter Spur, Wolfram Fischer (ed.): Georg Schlesinger und die Wissenschaft vom Fabrikbetrieb. Munich; Vienna: Carl Hanser 2000

[J. Z.]

Adolf Slaby (1849–1913)

Am 18. April 1849 wurde *Adolf Carl Heinrich Slaby* in Berlin als Sohn eines Buchbindermeisters geboren. Nach Absolvierung des Realgymnasiums studierte er von 1869 bis 1873 an der Berliner Gewerbeakademie Maschinenbau. Anschließend promovierte er mit einer mathematischen Arbeit an der Universität Jena.

1873 erhielt Slaby eine Stelle an der Kgl. Provinzial-Gewerbeschule in Potsdam, wo er Mathematik und Mechanik unterrichtete. Daneben begann er mit sehr praxisorientierten Arbeiten auf dem Gebiet des Maschinenbaus und befasste sich intensiv mit der Elektrotechnik. 1876 habilitierte er sich an der Gewerbeakademie für Theoretische Maschinenlehre. 1879 wurde er Mitbegründer des Berliner Elektrotechnischen Vereins. 1882 wurde Slaby auf die Professur für Theoretische Maschinenlehre und Elektrotechnik an die TH Berlin berufen. Zwei Jahre später übernahm er auch den Aufbau und die Leitung des elektrotechnischen Laboratoriums an der TH.

Slabys hohe gesellschaftliche Anerkennung beruhte wesentlich auf seiner rhetorischen Begabung und der Fähigkeit, komplizierte technische Zusammenhänge auch für Laien verständlich aufbereiten zu können. 1893 wurde Wilhelm II. (1859–1941) auf Slaby aufmerksam, nachdem dieser erfolgreiche Vorschläge für eine Abänderung der elektrischen Beleuchtung im weißen Saal des Berliner Schlosses unterbreitet hatte. In der Folge wurde Slaby zu einem persönlichen Berater des Kaisers in technischen Fragen. Wilhelm II. erschien regelmäßig, manchmal auch mit großem Gefolge aus Heer, Marine und Ministerien, zu Slabys Experimentalvorlesungen an der TH.

Seit Mitte der neunziger Jahre arbeitete Slaby fast ausschließlich auf dem Gebiet der drahtlosen Telegraphie. 1897 reiste er nach Großbritannien, um als Beobachter an den Funkversuchen von Guglielmo Marchese Marconi (1874–1937) teilzunehmen. Nach seiner Rückkehr baute

Adolf Carl Heinrich Slaby was born on 18 April 1849 in Berlin, the son of a bookbinder. Having finished school, he studied at the Berliner Gewerbeakademie engineering from 1869 to 1873. He went on to take a doctorate in mathematics at the University of Jena.

In 1873 Slaby took up a position at the Königliche Provinzial-Gewerbeschule (Royal Regional Technical Academy) in Potsdam, where he taught mathematics and mechanics. At the same time he started on extremely practically-oriented papers in the field of mechanical engineering, concentrating particularly on electrical engineering. In 1876 he became professor at the Gewerbeakademie for Theoretical Mechanics. He became a co-founder of the Berlin Society of Electrical Engineering in 1879. In 1882 Slaby became professor of Theoretical Mechanics and Electrical Engineering at the TH Berlin. Two years later he took over the building and directorship of the electrical engineering laboratory at the TH.

Slaby's high public profile was largely due to his rhetorical skills and his ability to explain complicated technical ideas to lay audiences. He came to the attention of Kaiser Wilhelm II (1859–1941) in 1893, after he had submitted a successful proposal to alter the electrical lighting in the White Room of the royal palace in Berlin. As a result of this, Slaby became the monarch's personal advisor on technical matters. Wilhelm II regularly turned up, sometimes with a retinue of military, naval and ministerial personnel, at Slaby's experiment-lectures at the TH. From the mid-1890s on, Slaby worked almost exclusively on wireless telegraphy. He travelled to Great Britain in 1897 to take part in the radio trials being carried out by Guglielmo Marchese Marconi (1874–1937). On his return to Berlin, Slaby made a reconstruction of Marconi's device and, together with his assistant Georg Graf von Arco (1869–1940), developed it further. In 1898 Georg Graf von Arco became an engineer at AEG, which marked

Slaby Marconis Gerät nach und entwickelt es anschließend mit seinem Assistenten Georg Graf von Arco (1869–1940) in Berlin weiter. 1898 wurde Graf Arco als Ingenieur bei der AEG angestellt, und damit begann die Zusammenarbeit von Slaby mit der AEG bei der industriellen Verwertung der „Funkentelegraphie".

Slaby entfaltete ausgedehnte Aktivitäten in den Berufs-, Fach- und Standesverbänden der Techniker, so war er 1893 Mitbegründer des Vereins Deutscher Elektrotechniker und dessen erster Vorsitzender sowie von 1906 bis 1908 Vorsitzender des Vereins Deutscher Ingenieure. Im Studienjahr 1894/95 war er Rektor der TH Berlin und wesentlich an der Vorbereitung einer Versammlung aller deutschen Technischen Hochschulen in Eisenach beteiligt, auf der die Forderung nach dem Promotionsrecht erneut erhoben wurde. Auf Slabys Einfluss ist es wohl zurück zu führen, dass Wilhelm II. den drei Technischen Hochschulen in Preußen anlässlich der Centenarfeier der TH Berlin 1899 das Promotionsrecht verlieh. Slabys Verdienste um die Anerkennung des Ingenieurstandes würdigte der VDI durch die Verleihung der Grashof-Gedenkmünze.

1898 hatte Wilhelm II. Slaby zum Mitglied des Herrenhauses auf Lebenszeit ernannt. Slaby war außerdem Mitglied der technischen Deputation für Gewerbe und Mitglied des Patentamtes. 1912 trat Slaby aus gesundheitlichen Gründen von seinen beruflichen und öffentlichen Ämtern zurück und verstarb am 6. April 1913 in Berlin.

Lit.: Karl-Heinz Manegold: Adolf Slaby, in W. Treue u. W. König: Berlinische Lebensbilder Bd. 6 – Techniker, Berlin: Colloquium Verlag 1990

[J. Z.]

the start of Slaby's collaboration with AEG on the industrial realisation of radio telegraphy.

Slaby was involved in a broad range of activities associated with professional and specialist organisations in the field of technology. He co-founded the German Association of Electrical Engineers and he served as the body's first chairman. He also headed the German Association of Engineers between 1906 and 1908. He was the rector of the TH Berlin in 1894/95 and he played an important role in the preparation of a congress of all German Technische Hochschulen in Eisenach, where the participants renewed the call for parity of status between degrees from traditional universities and from Technische Hochschulen. It was in large measure due to Slaby's influence that Wilhelm II granted degree-conferring status to the three Technische Hochschulen in Prussia on the occasion of the centenary of the TH Berlin in 1899. Slaby's services to the recognition of the engineering profession were recognised by the presentation of the Grashof Medal by the German Association of Engineers.

In 1898 Wilhelm II named Slaby a lifetime member of the royal household. Slaby was also a member of the Technical Deputation for Industry and a member of the patent office. In 1912, he retired from professional and public duties for health reasons and died in Berlin on 6 April 1913.

Lit.: Karl-Heinz Manegold: Adolf Slaby, in W. Treue u. W. König: Berlinische Lebensbilder Bd. 6 – Techniker, Berlin: Colloquium Verlag 1990

[J. Z.]

# Günter Spur (*1928)

Günter Spur wurde am 28. Oktober 1928 in Braunschweig geboren. Die Schulzeit war von den Kriegs- und Nachkriegswirren überschattet. 1948 begann er das Maschinenbaustudium an der Technischen Hochschule Braunschweig, und nach bestandener Diplomprüfung 1954 wurde er Konstrukteur bei der Werkzeugmaschinenfabrik Gildemeister in Bielefeld. 1956 wurde er zunächst Assistent, später Oberingenieur und Leiter des Versuchsfeldes am Institut für Werkzeugmaschinen und Fertigungstechnik der TH Braunschweig, wo er 1960 promovierte. 1961 kehrte Spur zur Firma Gildemeister zurück und übernahm dort die Konstruktionsleitung.

Am 1. Oktober 1965 wurde Spur auf den Lehrstuhl für Werkzeugmaschinen und Fertigungstechnik der Technischen Universität Berlin berufen und damit zum Direktor des gleichnamigen Instituts. Energisch machte er sich an die Erweiterung des Versuchsfeldes für Werkzeugmaschinen, 1968 konnte er eine neue Versuchshalle in Betrieb nehmen. Sofort griff Spur auch die am MIT begonnenen Arbeiten zur numerischen Steuerung von Werkzeugmaschinen auf. Spur hat maßgeblich zur Entwicklung der Programmiersprache EXAPT beigetragen, den Ausbau der CAD-Technologie forciert sowie neuartige Problemlösungen für den Einsatz von DNC- und CNC-Systemen erarbeitet. 1973 beschaffte Spur für das Institut für Werkzeugmaschinen und Fertigungstechnik (IWF) einen der ersten Industrieroboter und begann mit der Entwicklung von Robotersteuerungen. Daneben vernachlässigte Spur aber auch nicht die namensgebenden Forschungsfelder Werkzeugmaschinen, mit deren thermischen und dynamischen Verhalten er sich eingehend befasste, und Fertigungstechnik, wo insbesondere neue Werkstoffe und neue Bearbeitungsverfahren zahlreiche Forschungsthemen lieferten.

1976 gründete die Fraunhofer-Gesellschaft das von Spur angeregte Institut für Produktionsanlagen und Konstruktionstechnik. Durch einen Kooperationsvertrag mit der TU Berlin wurde es mit dem IWF zu

Günter Spur was born in Braunschweig on 28 October 1928. His years at school were overshadowed by the events of World War II and its aftermath. In 1948 he started to study engineering at the TH Braunschweig, where he took his degree in 1954. He then took up a post as a design engineer at the Gildemeister Tool-making Factory in Bielefeld. In 1956 he became an assistant, later chief engineer and then director of research at the Institute of Machine Tools and Production Engineering at the TH Braunschweig, where he took a doctorate in 1960. In 1961 he returned to work at Gildemeister as director of construction.

On 1 October 1965, Spur became professor and director at the Institute of Machine Tools and Production Engineering at the TU Berlin. He applied himself energetically to extending research in machine tools and opened a new research facility in 1968. At that point Spur started research into the numerical control of machine tools, an area that had already been started at the Massachusetts Institute of Technology. Spur made significant contributions to the development of the programming language EXAPT, he accelerated the upgrading of CAD technology and he also worked on the application of DNC and CNC systems. In 1973 he acquired one of the first industrial robots for his institute and he started to develop the field of robot control (robotics). Alongside these developments, Spur did not neglect the fields of research that gave his institute its name: he researched exhaustively the thermal and dynamic properties of machine tools, and in the field of production engineering a lot of research was conducted particularly into new materials and new processing procedures.

On Spur's motivation, the Fraunhofer Gesellschaft founded its Institute for Production Systems and Design Technology in 1976. The Fraunhofer Gesellschaft and the TU Berlin agreed to form a double institute by combining the former's Institute for Production Systems and Design

einem Doppelinstitut unter gemeinsamer Leitung von Spur verbunden. Im Neubau des Produktionstechnischen Zentrums (PTZ) in der Charlottenburger Pascalstraße, der 1986 fertig gestellt wurde, konnte Spur die beiden Institute auch räumlich integrieren. Er leitete das PTZ bis 1997 und machte es zu einem weltweit anerkannten Zentrum für die gesamte Produktionstechnik.

Spur organisierte zwischen 1975 und 1995 acht Produktionstechnische Kolloquien, die sich als internationale Foren sowohl des wissenschaftlichen Austausches wie des konstruktiven Gesprächs zwischen Forschung, Wirtschaft und Politik etablierten. Spur war in zahllosen Gremien und Ausschüssen tätig, unter anderem als Kurator der Physikalisch-Technischen Bundesanstalt und als Präsidialmitglied des Deutschen Instituts für Normung. Von 1991 bis 1996 war Spur Gründungsdirektor der Technischen Universität Cottbus. 1993 war er Gründungsmitglied der Berlin-Brandenburgischen Akademie der Wissenschaften und wurde Sekretar der technikwissenschaftlichen Klasse. Seine außerordentlichen Verdienste um die Ingenieurwissenschaft fand in unzähligen Preisen und Ehrenmitgliedschaften in zahlreichen Akademien Anerkennung. Mehrere in- und ausländische Universitäten verliehen Spur die Ehrendoktorwürde, und der VDI verlieh ihm 1991 die Grashof-Denkmünze.

Lit.: Frank-Lothar Krause, Eckart Uhlmann: Innovative Produktionstechnik, München, Wien: Carl Hanser 1998

[J. Z.]

Technology with the TU's Institute of Machine Tools and Production Engineering under the directorship of Spur. A new building known as the Production Technology Centre was completed in 1986 in Pascalstraße in Charlottenburg, Berlin, where Spur was able to combine the activities of both institutes. He remained director of the Production Technology Centre until 1997, making it an internationally recognised centre in the field of production technology.

Between 1975 and 1995, Spur organised eight colloquiums in production technology, which became international forums for the exchange of scientific knowledge as well as for constructive interaction between research, business and politics. Spur was a member of numerous bodies and committees, including the Physikalisch-Technische Bundesanstalt (Germany's national metrology institute), where he was curator, and he was a member of the board of the Deutsches Institut für Normung (German Institute for Standardization). Spur was the founding director of the University of Cottbus between 1991 and 1996. He became a founding member of the Berlin-Brandenburg Academy of Sciences in 1993 where he is secretary of the technical sciences division. Günter Spur's extraordinary achievements have been recognised with innumerable prizes and honorary memberships in many institutions. He holds honorary doctorates from many German and international universities and in 1991 the German Association of Engineers awarded him its prestigious Grashof medal.

Lit.: Frank-Lothar Krause, Eckart Uhlmann: Innovative Produktionstechnik, Munich, Vienna: Carl Hanser 1998

[J. Z.]

Iwan N. Stranski (1897–1979)

Als erster Sohn und drittes Kind eines bulgarischen Hofapothekers und einer Deutschbaltin wurde *Iwan Nicolá Stranski* am 2.1.1897 in Sofia geboren. Seine schweren Krankheiten von Kindheit an prägten ihn zeitlebens. Um sich selbst helfen zu können, entschloss er sich zum Medizinstudium, und um dies von einem höheren Niveau zu tun, studierte er erst einmal Chemie, zuerst in Wien, dann in Sofia, wo er 1922 das Diplom erwarb. Anschließend ging er nach Berlin an das physikalisch-chemische Institut der Universität, wo er bei Paul Günther (1892–1969) die Doktorarbeit über Röntgenspektralanalyse anfertigte und 1925 promoviert wurde. In Sofia wurde er 1926 zum o. Dozenten, 1929 zum a.o. und 1937 zum o. Professor ernannt. 1930/31 arbeitete Stranski als Rockefeller-Stipendiat am Institut für Physikalische Chemie der TH Berlin. Seitdem verband ihn eine lange wissenschaftliche Freundschaft mit Max Volmer (1885–1965). 1935/36 war er Abteilungsleiter am Physikalisch-technischen Institut des Urals in Swerdlowsk. 1941 kam Stranski endgültig nach Deutschland. 1941–1944 war er Gastprofessor an der Universität und TH in Breslau bei Rudolf Suhrmann (1895–1971), seit 1944 Wissenschaftliches Mitglied des Kaiser-Wilhelm-Instituts für physikalische Chemie und Elektrochemie in Berlin-Dahlem (jetzt Fritz-Haber-Institut) und seit 1953 dessen stellvertretender Direktor. Als Nachfolger von Max Volmer berief ihn die TU Berlin 1946 zum Ordinarius für physikalische Chemie. Als Honorarprofessor lehrte er bis 1963 auch an der FU Berlin. Stranski starb am 19.6.1979 in Sofia.

Seine Habilitationsarbeit über die Gleichrichterwirkung von Kristalldetektoren (1925) sowie eine Veröffentlichung von Erwin Madelung aus dem Jahre 1919 lenkten Stranskis Interesse auf die Probleme des Kristallwachstums und der Kristallauflösung. Erste Überlegungen führten ihn gleichzeitig mit und unabhängig von Walter Kossel (1888–1956) zu einem idealen Modell, dem „Kossel-Stranski-Kristall". Er schrieb der „Halbkristall-Lage" und der „mittleren Abtrennarbeit" zentrale Bedeutung zu. Weiterhin gab er eine kinetische Ableitung der Thomson-Gibbs-

Iwan Nicolá Stranski was born in Sofia, Bulgaria on 2 February 1897. He was the third child, and the first son, of a Bulgarian court pharmacist father and a mother of German-Baltic origin. As a child, he suffered from health problems, which persisted throughout his life. With the aim of making himself better, he decided to study medicine. He wanted to approach his goal from a higher level, so he embarked on studying chemistry, first in Vienna, Austria, then in Sofia, where he graduated in 1922. After that, he went to Berlin to the Physikalisch-chemische Institut of the University, where he did doctoral research on X-ray spectroscopic analysis under Paul Günther (1892–1969), receiving his doctorate in 1925. He taught in Sofia, where he started as a lecturer in 1926, eventually becoming professor in 1937. During 1930–31 Stranski won a Rockefeller scholarship at the Institut für Physikalische Chemie at the TH Berlin. It was then that he formed a long scientific friendship with Max Volmer (1885–1965, q.v.). In 1935–36 he was head of department at the Ural Institute of Physics and Mechanics in Sverdlovsk in the Soviet Union. In 1941 Stranski returned to Germany for good. During the years 1941–44 he was visiting professor under Rudolf Suhrmann (1895–1971) at the University and the TH Breslau (now Wroclaw, Poland). In 1944 he became a scientific fellow of the Kaiser-Wilhelm-Institut für physikalische Chemie und Elektrochemie (Kaiser Wilhelm Institute for Physical Chemistry and Electrochemistry, now the Fritz Haber Institute) in Dahlem, Berlin, becoming its associate director in 1953. The TU Berlin made him the successor to Max Volmer as professor of Physical Chemistry in 1946. He taught at the Free University of Berlin as an honorary professor until 1963. He died on 19 June 1979 in Sofia.

In his 1925 research paper for qualification to become a lecturer, he examined the rectifying effect of crystal detectors. This work, in conjunction with a 1919 paper by Erwin Madelung steered Stranski's interests towards the problems of crystal growth and crystal decay. His first deliberations led him simultaneously with, but independently of, Walter

schen Gleichung und mit den Gesetzen der Keimbildungshäufigkeit eine theoretische Ableitung der Ostwaldschen Stufenregel. Erwähnenswert sind Arbeiten über den Urotropinzerfall, die Polymorphie des Arsentrioxids, den Schmelzvorgang, die Tribolumineszenz und die Anwendung seiner theoretischen Erkenntnisse auf technische Prozesse der Stahlgewinnung. Mit Überlegungen zur Elektronenemission aus Kristallflächen hat er wesentlich zum Verständnis des Feldelektronenmikroskops von Erwin W. Müller (1911–1977) beigetragen.

Als Rektor setzte sich Stranski geschickt und erfolgreich für den Ausbau der TU Berlin ein, die ihm 1963 die Ehrensenator- und 1964 die Ehrendoktorwürde – eine von vielen – verlieh. Die Ernennung zum Auswärtigen Mitglied der Bulgarischen Akademie der Wissenschaften (1966) freute ihn besonders. Zwei Forschungsinstitute heißen nach ihm: das Iwan N.-Stranski-Institut der TU Berlin und das I. N.-Stranski-Institut für Metallurgie in Oberhausen.

Lit.: Kurt A. Becker und Jochen Block: Iwan N. Stranski, in: Max-Planck-Gesellschaft, Berichte und Mitteilungen, 1980, Heft 3; Rolf Lacmann: Iwan N. Stranski, in: Zeitschrift für Kristallographie 156 (1981), S. 167–175; R. Kaischew: On the history of the creation of the molecular kinetic theory of crystal growth, in: Journal of Crystal Growth 51 (1981), S. 643–650.

[M. E.]

Kossel (1888–1956) to an ideal model, the Kossel-Stranski Crystal. Stranski attributed a central significance to the ‚half-crystal position' and the concept of ‚mean separation work'. Further, he gave a kinetic derivation of the Gibbs-Thomson equation and he used the laws of nucleation to produce a theoretical derivation of Ostwald's step rule. Also worthy of mention are his works on the decomposition of Urotropin, the polymorphy of arsenic trioxide, melting processes, triboluminescence and the application of his theoretical knowledge to the technical processes of steel extraction. His work on the electron emissions of crystal surfaces made a significant contribution to Erwin W. Müller's (1911–1977) understanding of the field-emission microscope.

As rector, Stranski skilfully and successfully worked on the expansion of the TU Berlin, which made him an honorary senator in 1963 and awarded him one of his many honorary doctorates in 1964. He was especially pleased with his nomination as a foreign member of the Bulgarian Academy of Sciences in 1966. Two research institutes have been named after him: the Iwan N. Stranski Institute of the TU Berlin and the I. N. Stranski Institute for Metallurgy in Oberhausen.

Lit.: Kurt A. Becker und Jochen Block: Iwan N. Stranski, in: Max-Planck-Gesellschaft, Berichte und Mitteilungen, 1980, Heft 3; Rolf Lacmann: Iwan N. Stranski, in: Zeitschrift für Kristallographie 156 (1981), p. 167–175; R. Kaischew: On the history of the creation of the molecular kinetic theory of crystal growth, in: Journal of Crystal Growth 51 (1981), p. 643–650.

[M. E.]

Hugo Strunz (*1910)

Karl Hugo Strunz wurde am 24.2.1910 in Weiden (Oberpfalz) geboren, besuchte dort die Volks- und Realschule und anschließend die Oberrealschule in Regensburg. Hier trat er in den Naturwissenschaftlichen Verein Regensburg ein. 1929 begann er in München das Studium der Naturwissenschaften mit dem Schwerpunkt Mineralogie. Schon 1933 wurde er an der Ludwig Maximilians-Universität zum Dr. phil. promoviert und erwarb 1935 an der TH München den Dr. sc. techn. Dazwischen legte er das Referendar- und Assessorexamen für Naturwissenschaften ab, trat aber nicht in den höheren Schuldienst ein. Er ging als Forschungsstipendiat an das Institut von William Lawrence Bragg (1890–1971) nach Manchester und danach als Volontärassistent zu Paul Niggli (1888–1953) an die ETH Zürich, wurde 1937 Assistent von Paul Ramdohr (1890–1985) am Mineralogischen Museum in Berlin, habilitierte sich 1938 und wurde 1939 an der Friedrich Wilhelm-Universität Berlin zum Dozenten für Mineralogie und Petrographie ernannt. Nach dem Krieg kehrte Strunz nach Bayern zurück, erhielt einen Lehrauftrag für Mineralogie an der Philosophisch-Theologischen Hochschule in Regensburg und errichtete ein Mineralogisch-Geologisches Institut, das später zum Staatlichen Forschungsinstitut für Angewandte Mineralogie ausgebaut wurde. 1951 folgte Strunz dem Ruf auf ein Ordinariat für Mineralogie und Petrographie an die TU Berlin, nach seinen Worten der damals beste Lehrstuhl in Deutschland. Unter den erschwerten Bedingungen der Nachkriegszeit baute er in wenigen Jahren ein funktionsfähiges und sich ständig erweiterndes Institut auf, aus dem in den 27 Jahren seiner Tätigkeit mehr als 140 (von insgesamt ca. 220) Veröffentlichungen hervorgegangen sind. Nach der Emeritierung ging er wieder in seine Heimat zurück.

Als Kustos am Mineralogischen Museum wurde Strunz mit der Neugliederung betraut, die in eine Klassifikation aller Mineralien auf kristallchemischer Grundlage mündete. Die 1941 erstmals erschienenen „Mineralogischen Tabellen", seitdem mehrfach neu aufgelegt und in mehre Sprachen übersetzt, sind ein Standardwerk geworden. Folgerichtig

Karl Hugo Strunz was born in Weiden on 24 February 1910. He attended several schools in Regensburg. While there, he joined the Natural Sciences Society. In 1929 he started studying science in Munich, specialising in mineralogy. By 1933 he earned a PhD at the Ludwig Maximilians Universität and two years later became a doctor of science and technology at the TH Munich. Meanwhile he had completed two grades of exams to become a lecturer in science, but he did not take up a post in education. He won a research scholarship to the institute of William Lawrence Bragg (1890–1971) in Manchester, England and then he worked as an assistant to Paul Niggli (1888–1953) at the Eidgenössische Technische Hochschule (Swiss Federal Institute of Technology) in Zurich, Switzerland. In 1937 he became assistant to Paul Ramdohr (1890–1985) at the Mineralogisches Museum in Berlin. He gained the grade of professor in 1938 and found a post as teacher of mineralogy and petrography at the Friedrich Wilhelm Universität in Berlin in 1939. After World War II, Strunz returned to his native Bavaria in southern Germany and took a contract to teach mineralogy at the Philosophisch-Theologische Hochschule in Regensburg, where he set up a Mineralogisch-Geologisches Institut, which later was expanded to become the Staatliches Forschungsinstitut für angewandte Mineralogie (State Research Institute for Mineralogy). In 1951 Strunz came to the TU Berlin to become professor of mineralogy und petrography, which he said was the best professorship in Germany at the time. In the difficult postwar period he built up a fully-functioning and constantly-expanding institute in the space of a few years, which produced more than 140 publications (out of a total of 220) in the 27 years that he worked there. He returned to Bavaria after becoming professor emeritus.

As curator of the Mineralogical Museum, Strunz was charged with rearranging the collection according to a classification system based on crystal-chemical principles. His "Mineralogische Tabellen" (Mineralogical Tables), which was first published in 1941 and has been reissued many

befindet sich am Institut von Strunz auch die Internationale Zentralstelle für Mineraldatensammlung. So war er auch 1958–70 Vorsitzender der Mineral Data Commission in der International Mineralogical Association (IMA) und seit 1982 stellvertretend in diesem Amt. Als Mitverfasser von „Klockmanns Lehrbuch der Mineralogie" beeinflusste er die Ausbildung zahlreicher Studentengenerationen. Strunz war daneben auch an der Geschichte seines Faches interessiert. Neben mehreren Aufsätzen zeugt davon das 1970 erschienene Buch „Von der Bergakademie zur Technischen Universität Berlin 1770–1970".

Auf seinen zahlreichen Forschungsreisen in fast alle Kontinente entdeckte er zahlreiche neue Mineralien. Besondere Aufmerksamkeit widmete er den mineralogisch-geologischen Besonderheiten seiner oberpfälzischen Heimat. Von dort stammt auch ein 1957 beschriebenes Mangan-Eisen-Phosphat, das ihm zu Ehren Strunzit genannt wird; ein weiteres, Ferrostrunzit, trägt gleichfalls seinen Namen.

Strunz erhielt zahlreiche Würdigungen und Auszeichnungen. Er ist Mitglied mehrerer Akademien, bekleidete Vorstandsämter in internationalen und nationalen Fachgesellschaften. 1985 wurde ihm das Bundesverdienstkreuz erster Klasse verliehen.

Lit.: Werner Lieber: Hugo Strunz zum 90. Geburtstag am 24. Februar 2000, in: Lapis 25 (2000) Heft 2, S. 5–6; TU-Archiv.

[M. E.]

times and translated into many languages, is now regarded as a standard work. As a consequence, the Internationale Zentralstelle für Mineraldatensammlung (International Centre for Mineral Data Collection) is located at Strunz's institute. From 1958 to 1970 he headed the Mineral Data Commission of the International Mineralogical Association and he has been an associate director there since 1982. As a co-author of the mineralogy textbook "Klockmanns Lehrbuch der Mineralogie" he has influenced the education of several generations of students. Strunz has also demonstrated an interest in the history of his subject. He has published many articles on the theme and in 1970 he published a history of the 200 years of the mining academy of the TU Berlin, "Von der Bergakademie zur Technischen Universität Berlin 1770–1970".

On his numerous journeys to nearly all the world's continents he has discovered a large number of new minerals. He has devoted special attention to the mineralogical and geological particularities of his native Oberpfalz region in Bavaria. From this area comes a manganese-iron-phosphate, first described in 1957, which was named strunzite in his honour. Ferrostrunzite also bears his name.

Strunz has been awarded with many honours. He is a member of numerous academies and is on the boards of many national and international societies. In 1985 the West German state admitted him to the Order of Merit of the Federal Republic (first class).

Lit.: Werner Lieber: Hugo Strunz zum 90. Geburtstag am 24. Februar 2000, in: Lapis 25 (2000) Heft 2, p. 5–6; TU-Archive.

[M. E.]

Hans Heinz Stuckenschmidt wurde Musikhistoriker aus Zeitzeugenschaft, seine aufgezeichneten Tagesgeschichten der Musik verdichtete er zur Musikgeschichte des 20. Jahrhunderts.

Stuckenschmidt wurde am 1.11.1901 im damals deutschen Straßburg als Sohn eines Generalmajors und einer Pianistin geboren. Nach privatem Unterricht in Komposition, Klavier und Violine komponierte er seit seinem 18. Lebensjahr, gab Klavierkonzerte und leitete in den 1920er Jahren mit Josef Rufer (1893–1985) zwei Konzertzyklen. Ab 1920 arbeitete er als Berliner Musikkorrespondent bei der Prager Zeitschrift Bohemia. Als freier Musikschriftsteller in Hamburg, Wien, Paris, Berlin und Prag lernte er die führenden avantgardistischen Komponisten kennen, aber auch Maler, Literaten, Theaterleute, vor allem aus dem Kreis der Bauhauskünstler. Um nicht ein „gefürchteter Komponist und viel aufgeführter Kritiker" zu werden, gab er 1928 die eigene Musikausübung auf und widmete sich nun ausschließlich der Musikkritik, vor allem für die Berliner B.Z. am Mittag.

Die Teilnahme an Arnold Schönbergs (1874–1951) Analyse-Seminaren 1931–1933 prägte ihn nachhaltig und machte ihn zum engagierten Verfechter der Neuen Musik. Dafür von den Nationalsozialisten 1934 mit Schreibverbot belegt, ging er nach Prag und berichtete für das Prager Tagblatt und den Neuen Tag, wo ihn 1941 erneut das Schreibverbot traf. Nach dem Krieg kehrte er nach Berlin zurück und wurde Leiter der Abteilung Neue Musik beim amerikanischen RIAS, 1947 bei der renommierten, amerikanisch geleiteten Neuen Zeitung. 1947–1949 gab er mit Rufer die Zeitschrift „Stimmen" heraus.

1948 wurde Stuckenschmidt, der selbst nie studiert hatte, als Dozent an die TU Berlin berufen, die seit ihrer Wiedereröffnung nach dem Krieg ein Studium generale anbieten musste. 1953 zum Ordinarius

Hans Heinz Stuckenschmidt became a historian of the music to which he was a contemporary witness. He marshalled his daily reports of what he heard into a musical history of the 20th century.

He was born on 1 November 1901 in Straßburg (Strasbourg, France, then Germany), the son of a major-general and of a pianist. He received private lessons in composition, piano and violin and started composing from the age of 18. He gave piano concerts and directed two concert cycles with Josef Rufer (1893–1985) in the 1920s. From 1920 he worked as the Berlin music correspondent of the Bohemia newspaper, which was published in Prague. He worked as a freelance writer on music in Hamburg, Vienna, Paris, Berlin and Prague. He got to know the leading avantgarde composers of the era, as well as painters, literary figures, people working in theater, and above all the Bauhaus circle of artists. In order to avoid becoming a "dreaded composer and a much-performed critic," he gave up performing music in 1928 and devoted himself exclusively to musical criticism, especially for the Berlin newspaper BZ am Mittag.

He was deeply influenced by Arnold Schönberg's (1874–1951) analytic seminars of 1931–1933, making him a spirited advocate of the New Music movement. This led to his being banned from writing by the Nazis in 1934, whereupon he went to Prague and reported for the Tagblatt and Neuer Tag newspapers and was once again banned from writing in 1941. After World War II he returned to Berlin and became director of the department of new music at the RIAS (Radio in the American Sector), and in 1947 at the renowned American-led Neue Zeitung newspaper. In the period 1947–49 he and Rufer edited the magazine Stimmen.

In 1948, Stuckenschmidt, who had never studied, became a teacher at the TU Berlin, which since the end of the war was obliged to offer a general studies programme for science students. In 1953 he was made

für Musikgeschichte ernannt, beobachtete er weiter auf Reisen durch die ganze Welt das Musikgeschehen und berichtete darüber für so angesehene Zeitungen wie Die Welt, Frankfurter Allgemeine Zeitung, Neue Zürcher Zeitung und Stuttgarter Zeitung. Auch nach der Emeritierung 1967 blieb der brilliante Stilist publizistisch tätig.

Stuckenschmidt, der polyglotte Weltbürger, elegant und bissig-vornehm, war eine Schlüsselfigur der Berliner, ja der deutschen Musikszene. Mit Engagement kämpfte er für seine Überzeugung, der Gegensatz zu bürgerlicher Musik sei nicht die proletarische, sondern die weltbürgerliche Musik. Mit einem Buch über Schönberg 1951, für das er als erster dessen Nachlass auswerten konnte, begann eine ganze Reihe Musikerbiographien aus seiner Feder, über Igor Strawinsky, Boris Blacher, Johann Nepomuk David, Maurice Ravel, Ferruccio Busoni, Max Reger, vor allem 1974 das umfassende „Schönberg. Leben, Umwelt, Werk". Er machte die Leser mit der „Musik zwischen den beiden Kriegen", „Schöpfern der Neuen Musik" und „Oper in dieser Zeit" bekannt.

Stuckenschmidt wurde vor allem für sein publizistisches Werk mit dem Kritikerpreis der Salzburger Festspiele und dem Bundesverdienstkreuz geehrt, war Mitglied der Akademie der Künste Berlin und erhielt längst nach der Emeritierung seinen einzigen akademischen Grad, den Ehrendoktor der Universität Tübingen. Er starb am 15.8.1988 in seiner Wahlheimat Berlin.

Lit.: Stuckenschmidt, Hans Heinz: Ein Leben mit der Musik unserer Zeit – Zum Hören geboren, 1979. – Archiv der TU Berlin.

[B. E.]

professor of music history. He continued on journeys all over the world to monitor the music scene and he reported on it for such prestigious newspapers as Die Welt, Frankfurter Allgemeine Zeitung, Neue Zürcher Zeitung und Stuttgarter Zeitung. After becoming professor emeritus in 1967, he continued to write as a brilliant stylist.

Stuckenschmidt, a multilingual, elegant and sharp-tongued world citizen, was a key figure on the Berlin, indeed on the German, musical scene. He was a passionate proponent of the opinion that the opposite of bourgeois music was not proletarian music, rather cosmopolitan music. He published a book on Schönberg in 1951, making him the first to evaluate Schönberg's legacy. It was the first in a whole series of biographies of musicians, including Igor Stravinsky, Boris Blacher, Johann Nepomuk David, Maurice Ravel, Ferruccio Busoni and Max Reger. Particularly notable is his exhaustive 1974 book "Schönberg. Leben, Umwelt, Werk". He educated readers about the music of the interwar period in "Musik zwischen den beiden Kriegen", about the musicians behind the New Music in "Schöpfer der Neuen Musik", and about contemporary opera in "Oper in dieser Zeit".

It was primarily for his work as a writer that he was awarded the critics' prize of the Salzburger Festspiele music festival and the Order of Merit of the Federal Republic of Germany. He was a member of the Berlin Academy of Arts and, long after becoming professor emeritus, he achieved his only academic grade, that of honorary doctor at the University of Tübingen. He died on 15 August 1988 in his adopted home city of Berlin.

Lit.: Stuckenschmidt, Hans Heinz: Ein Leben mit der Musik unserer Zeit – Zum Hören geboren, 1979. – TU Berlin Archive.

[B. E.]

Herbert Sukopp (*1930)

Der am 6.11.1930 in Berlin geborene *Herbert Sukopp* besuchte zunächst das Leibniz-Gymnasium, 1945 das renommierte Berlinische Gymnasium zum Grauen Kloster. 1949 studierte er an der Pädagogischen Hochschule (PH) Berlin mit dem Wahlfach Biologie und legte 1953 das Lehramtsexamen ab. Als Hilfsassistent an der PH studierte er an der Freien Universität Berlin (FU) Botanik, Geologie und Soziologie. Er wurde dort 1957 am Institut für Systematische Botanik und Pflanzengeographie Wissenschaftliche Hilfskraft und 1958 mit einer Arbeit über die Berliner Moore zum Dr. rer. nat. promoviert.

Sukopp war zunächst Assistent am Institut für Angewandte Botanik, an dem er sich 1968 habilitierte und 1969 eine Professur bekam. Mit seinen Kollegen führte er dann drei Institute zum Institut für Ökologie zusammen und übernahm 1974 die Leitung des neugegründeten Fachgebiets Ökosystemforschung und Vegetationskunde. Bis nach der Emeritierung 1996 betreute er Diplomanden und Doktoranden auch von der FU.

Sukopp gilt als ein „Vater der Stadtökologie". Die isolierte Lage West-Berlins lenkte sein Interesse auf den wechselseitigen Einfluss von Mensch und Natur, das Ausmaß und die Folgen menschlicher Eingriffe auf Pflanzen, Tiere, Klima, Böden und Gewässer. Seine Untersuchungen führten zur ersten „Roten Liste" der Farn- und Blütenpflanzen für die Bundesrepublik Deutschland 1974 und 1977/78 für Europa. Seine ersten Kartierungen städtischer Biotope wurden inzwischen wissenschaftlicher Standard. Sukopp entwickelte in der Ökosystemforschung eine Untersuchungssystematik, die den Umweltverträglichkeitsstudien zugrunde liegt und sich etwa im Berliner Umweltatlas niederschlug. Das Konzept des „Natürlichkeitsgrades" (Jaakko Jalas, 1955) baute er für besiedelte Gebiete aus.

Sukopp verflocht stets Grundlagenarbeit mit angewandten Fragen und setzte auf die Diskussion zwischen Wissenschaftlern, Planern, Poli-

Herbert Sukopp was born on 6 November 1930 in Berlin. He entered the Leibniz-Gymnasium, and from 1945 attended the prestigious Berlinische Gymnasium zum Grauen Kloster. 1949 he entered the Pädagogische Hochschule in Berlin, where he specialised in biology. He qualified as a teacher in 1953. While he worked as an assistant at the Pädagogische Hochschule he also studied botany, geology and sociology at the Free University of Berlin. He took work as a graduate student in 1957 at the University's Institut für Systematische Botanik und Pflanzengeographie and he earned his doctorate in 1958 with a dissertation on the marshes of Berlin.

He was an assistant at the Institut für angewandte Botanik, where he qualified as a lecturer in 1968 and as a professor in 1969. He and his colleagues then drew together three institutes to create the new Institut für Ökologie. In 1974 Sukopp took over as head of the newly-founded subject Ökosystemforschung und Vegetationskunde. He continued to supervise master's degree and doctoral students from the Free University, even after becoming professor emeritus in 1996.

Sukopp is regarded as father of urban ecology. The isolated situation of West Berlin focussed his attention on the interactive influences of humans and nature and on the extent and the consequences of human intervention on plants, animals, climate, soil and water. His investigations led in 1974 to the first ‚red list' (list of endangered species) of ferns and flowering plants in West Germany and a ‚red list' for Europe in 1977/78. His first mappings of urban biotopes became the scientific standard. He developed an investigative approach in the field of ecosystem research that has provided a basis for studies of environmental sensitivity and is reflected to some degree in the Berliner Umweltatlas. He expanded the concept of "Natürlichkeitsgrad" ('degree of naturalness', Jaakko Jalas, 1955) to apply to built-up areas.

tikern und Bevölkerung. In der Überzeugung, der Naturschutz in Städten müsse den Menschen dienen und daher vor allem die Klima- und Lufthygiene beachten, half er, in Berlin schädliche Eingriffe des Menschen zu verhindern (Kraftwerk Oberhavel) oder abzumildern (Autobahnzubringer durch den Tegeler Forst), aber auch aufgrund seiner Untersuchung der Ufervegetation der Havel, die zum Röhrichtschutzgesetz führte, die Renaturierung der Ufer durchzusetzen. Mit freundlicher, dialogfähiger Sachlichkeit und wohlüberlegter, beharrlicher Argumentation entfaltete er große Wirkung für den Menschen und seine Umwelt.

In zahlreichen Gremien für Naturschutz, Landschaftspflege und Umweltfragen war er ab 1974 zunächst in Berlin, dann auch auf Bundesebene und beim Europarat, im IUCN und für die UNESCO beratend oder leitend ehrenamtlich tätig. Seine wissenschaftlichen Verdienste wurden mit dem Ehrendoktorat der TU München gewürdigt, die Mitgestaltung politischer und gesellschaftlicher Prozesse im Sinne des Naturschutzes wurde 1983 mit dem Bundesverdienstkreuz, 1987 mit dem Verdienstorden des Landes Berlin, Preisen und Ehrenmitgliedschaften ausgezeichnet. Sukopp wurde zudem als Mitglied in die Deutsche Akademie für Städtebau und Landesplanung und 1995 in die Berlin-Brandenburgische Akademie der Wissenschaften aufgenommen.

Lit.: Dynamik und Konstanz. Festschrift für Herbert Sukopp, zusgest. von Ingo Kowarik ..., Bonn-Bad Godesberg 1995 (= Schriftenreihe für Vegetationskunde; Bd. 27).

[B. E.]

Herbert Sukopp (*1930)

Sukopp always interwove fundamental principles with real-life issues and he set great store by discussions between scientists, planners, politicians and local population. He was convinced that nature protection in cities must have the interests of people at heart, above all it must promote a good climate and clean air. Thus he helped either to prevent damaging human impact (such as the Oberhavel power station) or to reduce its effects (such as the motorway feeder road through Tegeler Forst forest). His survey of the vegetation on the banks of the river Havel led to a law for the protection of reeds and so to the reclaiming of the riverbanks for nature. Sukopp's friendly, conversational and objective approach and his well-considered and perseverant mode of putting his ideas across have had a major impact of the way people live and on the environment.

Starting in Berlin in 1974, he has been either an advisor or in a leading role, and always in a voluntary capacity, in various bodies devoted to nature protection, landscape preservation and environmental issues; he has also worked in similar bodies at the federal level and in the European Council, at the World Conservation Union and for UNESCO. The TH Munich has honoured his scientific achievements with an honorary doctorate. His combination of the political and the scientific in the interests of natural protection earned him the Order of Merit of the Federal Republic of Germany in 1983, and in 1987 the order of merit of the federal state of Berlin, as well as other prizes and honorary memberships. Sukopp is also a member of the Deutsche Akademie für Städtebau und Landesplanung and he was inducted into the Berlin-Brandenburgische Akademie der Wissenschaften (Berlin-Brandenburg Academy of Sciences) in 1995.

Lit.: Dynamik und Konstanz. Festschrift für Herbert Sukopp, edited by Ingo Kowarik ..., Bonn-Bad Godesberg 1995 (= Schriftenreihe für Vegetationskunde; Bd. 27).

[B .E.]

Die Geschichte der Wissenschaft sei die Wissenschaft selbst, heißt es. Zwar möchte nicht jeder zustimmen, doch wenn sich jemand wie *István Szabó* sowohl als Wissenschaftler und als auch als Wissenschaftshistoriker auszeichnet, wählt er bemerkenswerter jene Worte aus Goethes *Farbenlehre*, um den Zweck und den Nutzen seiner geschichtlichen Forschungen zu begründen.

István Szabó wurde am 13. Dezember 1906 im ungarischen Orosháza geboren; nachdem er in Ungarn aufgewachsen war, ging er 1926 nach Berlin, um an der Technischen Hochschule Physik zu studieren; von 1930 bis 1933 arbeitete er im Forschungslabor der Firma Osram und beendete 1934 sein Studium als Diplom-Ingenieur. Anschließend war er zunächst wieder in der Industrie tätig und ab 1940 an der TH als Assistent des Mathematikers Werner Schmeidler, bei dem er 1943 zum Dr.-Ing. promovierte; 1946 wurde er Mitglied der Fakultät für Allgemeine Ingenieurwissenschaften, habilitierte sich 1947 für das Lehrgebiet Mathematik und wurde 1948 zum Professor der Mechanik berufen, welchen Lehrstuhl er bis zu seiner Emeritierung im Jahre 1975 innehatte.

Als Hochschullehrer genoss István Szabó einen glänzenden Ruf, denn er besaß in ungewöhnlichem Maße die Begabung, schwierigen Wissensstoff nicht nur zu selbst zu meistern, sondern diesen dann auch seinen Hörern verständlich darzulegen. Dank seiner hervorragenden mathematischen Kenntnisse brachte er alle Voraussetzungen mit, um die Mechanik frühzeitig auf ein neues mathematisches Niveau zu heben, welches sich schon bald als unverzichtbar erweisen sollte. Hierzu entwickelte und benutzte er mathematische Hilfsmittel, die bis dahin eigentlich als zu schwierig gegolten hatten, um sie Studenten der Ingenieurwissenschaften zuzumuten. Doch es bewahrheitete sich, dass der Mensch mit seinen Aufgaben wächst, denn István Szabó schreckte damit die Studenten keinesfalls ab; die Hörerzahlen belegen vielmehr, dass sie offensichtlich erfreut darüber waren, in der Mechanik auf einem hohen mathematischen Niveau ausgebildet zu werden. Seine didaktischen Fähigkeiten

The history of science is science itself, it is said. While not everyone might agree, when somebody like *István Szabó* excels both as a scientist and as a scientific historian, then he can use the notable words above from Goethe's *Farbenlehre* (theory of colour) to justify the purpose and value of his historical research.

István Szabó was born in Orosháza, Hungary on 13 December 1906. After growing up in Hungary he left for Berlin in 1926 to study physics at the Technische Hochschule. From 1930 to 1933 he worked at Osram's research laboratory, completing his studies as a graduate engineer in 1934. Subsequently he was employed in industry again, and from 1940 was employed at the Technische Hochschule as an assistant to the mathematician Werner Schmeidler. He was awarded a doctorate (Dr.-Ing.) in 1943. In 1946, he became a member of the faculty for General Engineering Sciences in 1947. He then qualified as a university lecturer for mathematics and was called upon to be Professor of Mechanics in 1948, a chair he was to hold until he was given emeritus status in 1975.

István Szabó enjoyed a shining reputation as a university lecturer, for he was highly gifted – not only with the ability to master difficult intellectual material but also to make it understandable for his listeners. Thanks to his excellent mathematical knowledge, he had all the qualities needed to take mechanics rapidly to a new mathematical level, a level that soon proved itself to be indispensable. To this end he developed and used mathematical aids that, up to that time, had been regarded as being far too difficult for engineering students to be expected to use. Yet it has been demonstrated that people grow with their responsibilities, for István Szabó did not frighten off his students at all. Instead, the numbers attending showed that students were clearly happy to be taught in mechanics at such a high mathematical level. István Szabó was also able to confirm his didactic skills in his books on technical mechanics. The influence of Georg Hamel, whom István Szabó admired, is unmistakable

bewies István Szabó zudem in seinen Büchern über Technische Mechanik;
hinsichtlich des Inhaltes ist der Einfluss des von István Szabó geschätz-
ten Georg Hamel unverkennbar, sodass die Bücher den Untertitel *Hamel
für Ingenieure* tragen könnten. Sie fanden weite Verbreitung nicht nur
bei seinen Hörern und den Studenten der TU Berlin, sondern auch bei
Studenten anderer Universitäten – und István Szabó wird weiterhin Leser
finden, denn die *Höhere Technische Mechanik* ist erfreulicherweise jüngst
als ein Klassiker der Technik neu aufgelegt worden.

Wie bei den anderen hier vorgestellten Personen aus dem mathematischen Bereich fällt auch bei István Szabó die Weite des Horizontes
auf; so hat er sich – längst selbst ein Klassiker geworden – den Klassikern
der Mechanik vor ihm zugewandt und eine *Geschichte der mechanischen
Prinzipien und ihrer wichtigsten Anwendungen* geschrieben, welche sorg-
fältige Untersuchung durch gründliches und umfangreiches Studium
der Originalquellen besticht. Hat man dieses Buch selbst so sorgfältig
gelesen, wird sich ein weiterer von István Szabó gewählter Leitspruch
bewahrheiten: „Ich habe die vorderste Linie rasch erreicht, weil ich die
Meister und nicht ihre Schüler studiert habe."

---

*Lit.: Einführung in die Technische Mechanik. 8. Auflage Berlin u. a. 1975. – Höhere
Technische Mechanik. 5. Auflage Berlin u. a. 1977. – Geschichte der mechanischen
Prinzipien und ihrer wichtigsten Anwendungen. Basel u. a. 1979.*

[F. H.]

in these books, so that they could also have borne the subheading *Hamel für Ingenieure* (Hamel for engineers). They not only found eager readers among his audience and the students at the Technische Universität Berlin but also among students at other universities. István Szabó will continue to find new readerships as, happily, *Höhere Technische Mechanik* (Higher Technical Mechanics) has recently been reprinted as a technological classic.

Just as with the other people from the world of mathematics introduced here, the scale of István Szabó's intellectual horizon stands out. He turned his attention towards the earlier classics of mechanics (although he had long been one himself) and wrote *Eine Geschichte der mechanischen Prinzipien und ihrer wichtigsten Anwendungen* (A history of mechanical principles and their most important applications), a painstaking investigation that impresses through its thorough and extensive study of the original sources. After carefully reading this book, another motto chosen by István Szabó proves true: "I got to the very top because I studied the masters and not their students."

Lit.: Einführung in die Technische Mechanik. 8th edition Berlin, etc. 1975. – Höhere Technische Mechanik. 5th edition Berlin, etc. 1977. – Geschichte der mechanischen Prinzipien und ihrer wichtigsten Anwendungen. Basel, etc. 1979.

[F. H.]

„Und ich lebe von Licht", singt der Mensch, den man laut Heinrich Zille auch mit einer Wohnung erschlagen könne, mit Wohnungen, wie sie Zille tagtäglich im Berlin der Mietskasernen und deren berühmt-berüchtigten düsteren Hinterhöfen vor Augen hatte. Es galt zu handeln, während der Weimarer Republik wurde gehandelt, und Siedlungen entstanden, die dank der beteiligten Architekten wie *Bruno Taut* nur berühmt sind.

Geboren wurde Bruno Taut am 4. Mai 1880 in Königsberg. Seit 1897 besuchte er die Baugewerbeschule; nachdem er diese erfolgreich abgeschlossen hatte, ging er um die Jahrhundertwende aus Königsberg fort und ließ sich nach wechselnden Aufenthaltsorten 1908 in Berlin nieder. Hier lernte er Künstler wie den Dichter Paul Scheerbart kennen, dessen Visionen einer Glasarchitektur ihn beeinflussten, und zudem begann er gegen 1917, eigene Architekturvisionen wie die Bilderreihen *Alpine Architektur* mit einem Dom aus Glas und *Die Auflösung der Städte* zu entwerfen; hierzu war er durch Kropotkin angeregt worden, der mit Hilfe der Wissenschaft Stadt und Land miteinander versöhnen wollte. Handelte es sich dabei noch um Utopien, so wandte sich Bruno Taut dann ganz im Sinne des „hier, oder nirgends ist Amerika" einer Verbesserung der Wohnsituation in der Wirklichkeit zu und arbeitete an Siedlungen des gemeinnützigen Wohnungsbaues mit; er entwarf ungefähr zehntausend Wohnungen, seine bekanntesten und gelungensten Bauten sind die Waldsiedlung in Zehlendorf und die Hufeisensiedlung in Britz. Luft, Sonne und Grün wurden großzügig in die Wohnanlagen miteinbezogen, weshalb Bruno Taut rückblickend schrieb, damals sei die Sonne für den Städtebau entdeckt worden.

Doch in Deutschland sollte die Sonne erlöschen. Ein Licht auf die politische Lage der Zeit werfen schon die Begleitumstände der im Jahre 1930 erfolgten Berufung Bruno Tauts an die TH Berlin als Honorarprofessor, welche auf harten Widerstand der Fakultät stieß, weil er offen

"And I live from light" goes the song. Berliner artist Heinrich Zille also saw things that way. In his view, you could beat people to death with an appartment, or at least with the kind of accomodation he saw every day in his Berlin of "rented barracks" with their famous and infamous back yards. Something had to be done and it was done during Germany's "Weimar Republic" following the First World War when housing estates were built that gained fame but not notoriety, thanks to the architects involved, like *Bruno Taut.*

Bruno Taut was born in Königsberg (now Kaliningrad, Russian Federation) on 4 May 1880. He attended the Baugewerbeschule (Construction Trades School) from 1897. After successfully completing his studies there, he left Königsberg at the turn of the century and settled in Berlin in 1908, after living at a succession of different places. It was in Berlin that he got to know artists like poet Paul Scheerbart, whose visions of an architecture of glass were to influence him, and in 1917 he also began to draw up his own architectural visions like the Alpine Architecture series of pictures with a glass cathedral, and Die Auflösung der Städte (The dispersal of the cities). He was also excited by Kropotkin, who wanted to reconcile the city and the countryside with the aid of science. If these were still mere utopias, Taut then turned his attention to improving actual living situations and worked to design estates for non-profit-making housing associations. He designed around ten thousand apartments, his most important and successful works being the Waldsiedlung (Forest Estate) in Zehlendorf and the Hufeisensiedlung (Horse-Shoe Estate) in Britz (both Berlin). These residential estates made much use of the air, the sun and green vegetation, which is why Taut wrote, in retrospect, that the use of the sun in urban development was discovered at that time.

However, the sun was beginning to go out in Germany. The circumstances surrounding the offer of an appointment to Taut as honorary professor at the Technische Hochschule Berlin in 1930 (which met with

mit der Arbeiterbewegung sympathisierte. Als Hochschullehrer des Wohnungsbaues und Siedlungswesens wollte Taut seinen Hörern nicht bloßes Wissen vermitteln, vielmehr sollten sie sich ihres eigenen Verstandes bedienen und Wohnungen nicht einfach nach einem bestimmten starren Schema entwerfen. Insbesondere lag ihm am Herzen, dass sich die angehenden Architekten der hohen Verantwortung bewusst würden, die mit dem Wohnungsbau verbunden sei; hierzu schlug er neue Wege ein und ließ die Studenten Untersuchungen durchführen, damit sie Kenntnisse über die tatsächlichen Wohnverhältnisse und Lebensbedingungen in den Arbeitervierteln erhielten oder damit sie lernten, wie Siedlungen anzulegen und auszustatten seien.

Bruno Tauts Tätigkeit an der TH Berlin endete 1932, denn er ging nach Moskau, um dort als Architekt zu arbeiten. Jedoch erfüllten sich seine Hoffnungen nicht, und nachdem er vor Ort in der Sowjetunion ernüchtert worden war, kehrte er – die Gefahr verkennend – Ende Februar 1933 nach Berlin zurück. Dank einer Warnung entkam er seiner drohenden Verhaftung und war nun heimatlos; seine Flucht führte ihn zunächst nach Japan und 1936 in die Türkei, wo Bruno Taut am Heiligabend 1938 verstarb. Seine *Architekturlehre* erschien postum erst auf Türkisch, dann auf Japanisch und Jahrzehnte später schließlich auch auf Deutsch.

Lit.: Der Weltbaumeister. Architekturschauspiel für symphonische Musik. Hagen 1920. – Die neue Wohnung. Die Frau als Schöpferin. Leipzig 1924. – Die neue Baukunst in Europa und Amerika. Stuttgart 1929. – Architekturlehre. Hamburg und Berlin 1977.

[F. H.]

bitter resistance from the faculty because of his open sympathy for the labour movement) throws light on the contemporary political situation. As university lecturer of housing construction and urban development, Taut did not just merely disseminate knowledge to his audience. Instead, his students had to use their own intelligence and not merely design apartments according to a specific and rigid pattern. In particular, he took it to heart that the prospective architects became aware of the high level of responsibility involved in building housing. To this end, he led in new directions and let the students conduct investigations to find out about the real housing and living conditions in working-class areas or learn how to develop and furnish housing estates.

Bruno Taut's activity at the TH Berlin came to an end in 1932, when he left for Moscow to work as an architect. However, his hopes remained unfulfilled, and after being brought down to earth in the USSR he returned to Berlin in late February 1933 – unconscious of the danger that awaited him. He was able to evade the threat of arrest thanks to a warning he received, but he was now without a home. His escape took him first to Japan and then to Turkey in 1936, where he died on Christmas Eve 1938. His Architekturlehre (Study of architecture) was published postumously – first in Turkish, then in Japanese and finally, decades later, in German.

Lit.: Der Weltbaumeister. Architekturschauspiel für symphonische Musik Hagen 1920. – Die neue Wohnung. Die Frau als Schöpferin. Leipzig 1924. – Die neue Baukunst in Europa und Amerika. Stuttgart 1929. – Architekturlehre Hamburg and Berlin 1977.

[F. H.]

Oswald Mathias Ungers (*1926)

Urbilder und Geometrie findet, wer das Schöne in Schlichtheit liebt; nun entwirft er Gebäude, und lässt er in immer neuen Gestalten jene Bilder erscheinen, dann gelingt *Oswald Mathias Ungers* das Zusammenfallen der Gegensätze Goethe und Geometrie.

Oswald M. Ungers wurde am 12. Juli 1926 in Kaiseresch in der Eifel geboren, studierte von 1947 bis 1950 Architektur an der TH Karlsruhe und eröffnete nach dem erfolgreichen Abschluss seines Studiums ein Architekturbüro in Köln und Berlin. Von 1963 bis 1969 lehrte er als Professor an der TU Berlin, danach an der Cornell University Ithaca, deren Chairman er von 1975 bis 1986 war; zwischendurch hatte er noch andere Professuren inne, so an der Harvard University, und anschließend war er von 1986 bis 1990 Professor an der Kunstakademie Düsseldorf.

Oswald M. Ungers lässt sich von dem Wahlspruch „Varietà in unità" leiten, diese Verschiedenheit in der Einheit prägt sein morphologisches Denken und Handeln. Zunächst sind Urformen zu erkennen, welche dann immer wieder anders umgestaltet werden; die Verschiedenheit ist folglich keine beliebige Unterschiedlichkeit, vielmehr handelt es sich um mannigfaltige Abwandlungen dieser allen gemeinsamen Urform, dieser Idee, und so besteht die Einheit in dem zu Grunde liegenden, nur gedachten Archetypus, aber nicht in den gebauten Erscheinungen, welche sonst Stereotypen wären. Ungers führt das Beispiel der Verwandlung einer Pflanze an, und in der Tat erinnern seine Überlegungen an Goethe: „Alle Gestalten sind ähnlich, und keine gleichet der andern; und so deutet der Chor auf ein geheimes Gesetz."

Das Gesetz entdeckt Ungers in der Geometrie, und es sind einfache Figuren, wie Gerade, Dreieck, Quadrat, Kreis, Würfel, Zylinder, die den Ausgangspunkt des architektonischen Entwurfes bilden und Größe und Gestalt der Gebäude samt ihrer Räume festlegen. Dieser Entwurf erfolgt

A lover of beauty in simplicity will find his own archetypes and geometry. Designing buildings, he will let these images appear in ever new forms, for *Oswald Mathias* Ungers was able to break down the contradictions between Goethe and geometry.

Oswald M. Ungers was born at Kaiseresch in the Eifel region on 12 July 1926, studied architecture at the Technische Hochschule in Karlsruhe from 1947 to 1950 and, after successfully completing his studies, opened an architect's office in Cologne and Berlin. He taught as a professor at the Technische Universität Berlin from 1963 to 1969, then at Cornell University in Ithaca, of which he was Chairman from 1975 to 1986. In between times, he held other professorships, including one at Harvard University, before becoming Professor at the *Kunstakademie* (academy of arts) in Dusseldorf from 1986 to 1990.

Oswald M. Ungers was guided by the motto "variety in unity", which shaped his morphological thoughts and deeds. First, one identifies archetypes that are reshaped differently each time – so that the differences arising are not just mere dissimilarities but rather diverse variations on the common archetype and idea. By these means, a unity is achieved based on the fundamental, conceived archetype, though not in the buildings created, which would otherwise just be stereotypes. Ungers cites the example of the growth of a plant, and his ideas do indeed remind us of Goethe's: "All the designs are similar – yet none is the same. The chorus suggests a common law."

Ungers discovered this law in geometry, and in the simple figures, such as straight lines, triangles, squares, circles, cubes, and cylinders that form the starting point of architectural design, determining the size and shape of the buildings and their rooms. This design does not arise arbitrarily outside of time and space, and to be more precise, Ungers stresses that architecture is tied to a location, to its history and spirit, and must

nicht willkürlich jenseits von Zeit und Raum, vielmehr betont Ungers, dass die Architektur an den Ort, an dessen Geschichte und Geist gebunden sei und sich in das Vorhandene einzuordnen habe; die jeweiligen Umstände mit all ihren Unzulänglichkeiten seien hinzunehmen, denn die Architektur könne sich nicht der geschichtlichen Wirklichkeit entziehen. Jedoch sollen einst geschaffene Gestalten nicht auf ewig unverändert weitergeführt, sondern umgeschaffen werden, sodass etwas Neues im Sinne der Verwandlungslehre entsteht, indem nämlich Altes umgestaltet wird; der Ort soll nicht zu Tode erstarren, sondern beweglich und damit lebendig bleiben, indem er durch eine andere Sichtweise belebt und neu erlebt wird.

Über bloße Absichten hinaus ist es Ungers vor allem gelungen, seine Gedanken über Architektur in einem eigenen Stil überzeugend umzusetzen. Zwei der vielen beeindruckenden Beispiele sind das Deutsche Architekturmuseum in Frankfurt am Main und die Residenz des Deutschen Botschafters in Washington, die gemäß Ungers' Absicht weitaus mehr als einen bloßen Funktionsbau darstellen soll, und zwar einen „Spiegel deutscher Baukultur" als eine Visitenkarte, um sich dem Gastland vorzustellen.

Es sind sowohl die Betrachtungen über Architektur als auch die Architektur selbst, die einen unwillkürlich wünschen lassen, dass ein Architekt wie Oswald Mathias Ungers den Schlossplatz in Berlin gestalten möge.

Lit.: Architektur 1951–1990. Stuttgart 1991. – Bauten und Projekte 1991–1998. Stuttgart 1998. – 10 Kapitel über Architektur. Ein visueller Traktat. Köln 1999.

[F. H.]

be integrated in what is there. The differing circumstances have to be accepted (with all their shortcomings) because architecture cannot be detached from historical reality. Nevertheless, forms that have once been created cannot stay unchanged for ever but instead have to be restructured so that something new arises in the sense of transformation theory, and that the old is re-fashioned. The location should not become motionless and dead but movable and living, through being brought to life and re-lived via a different perception.

Transcending mere intentions, Unger above all succeeded in being able to put his ideas on architecture into practice convincingly and in his own style. Two of many impressive examples are the German Architectural Museum in Frankfurt am Main, and the residence of the German Ambassador in Washington, which it was Ungers' intention to represent more than just a mere functional building by being a "Mirror of German building culture" and a visiting card to present to the host country.

It is both his observations on the subject of architecture and his architecture itself that makes one instinctively wish that an architect like Oswald Mathias Ungers would re-shape Berlin's Schlossplatz, the site of the former City Palace.

Lit.: Architektur 1951–1990. Stuttgart 1991. – Bauten und Projekte 1991–1998 Stuttgart 1998. – 10 Kapitel über Architektur. Ein visueller Traktat. Cologne 1999.

[F. H.]

Max Volmer wurde am 3.5.1885 in Hilden geboren, genoss eine gediegene humanistische Bildung, widmete sich der Musik und Literatur sowie der Biologie. Als erfolgreicher Entomologe und Orchideenzüchter blieb er zeitlebens den biologischen Neigungen treu. Er begann 1905 in Marburg das Chemiestudium und schloss sich dort Karl Schaum (1872–1947) an, der Volmers Interesse auf die Photochemie lenkte. 1908 folgte er Schaum nach Leipzig an das von Max Le Blanc (1865–1943) geleitete physikalisch-chemische Institut und wurde 1910 promoviert. Nach kurzem Militärdienst wurde Volmer Assistent am Leipziger Institut und habilitierte sich 1913 für anorganische und physikalische Chemie.

Im August 1914 wurde er eingezogen und 1916–18 als „Gasschutzoffizier" an das physikalisch-chemische Institut der Berliner Universität abgeordnet. Volmers „kriegswichtige Forschung" war marginal, nicht jedoch die Nähe zu Walther Nernst (1864–1941), dessen physikalische Denkweise ihn beeinflusste und darin bestärkte, bei allen wissenschaftlichen Problemen das Grundsätzliche zu suchen. An Nernsts Institut arbeitete er auch mit dem Physiker Otto Stern (1888–1969) zusammen. Die „Stern-Volmer-Gleichung" für die Intensität des Fluoreszenzlichts ist noch heute gültig. Dabei entwickelte er eine patentierte Quecksilberdampfstrahlpumpe. Ein guter Blick für die technische Umsetzung und Wirtschaftlichkeit seiner Forschung zeichnete ihn aus.

Nach kurzem Zwischenspiel als Leiter des wissenschaftlichen Laboratoriums der Auer-Gesellschaft nahm er 1920 den Ruf an die Hamburgische Universität an, wechselte jedoch 1922 auf die Stelle eines o. Professors und Direktors des Instituts für physikalische Chemie und Elektrochemie der TH Berlin. Hauptarbeitsgebiete waren die Phasenbildung und das Kristallwachstum sowie die Überspannung bei Elektrodenvorgängen. Nach 1933 war er verschiedenen Repressalien ausgesetzt, die mit der Verweigerung der Bestätigung seiner Wahl in die Preußische Akademie

Max Volmer was born in Hilden on 3 May 1885. He enjoyed a good education, devoting himself to music, literature and biology. He was a successful entomologist and orchid breeder and he remained true to his aptitude for biology throughout his life. He started studying chemistry in Marburg in 1905 and came into contact with Karl Schaum (1872–1947), who steered Volmer's interests in the direction of photochemistry. In 1908 he followed Schaum to the Physikalisch-chemisches Institut in Leipzig, which was under the direction of Max Le Blanc (1865–1943). Volmer took his doctorate in 1910. After a short period of military service, he became an assistant at the institute in Leipzig and qualified to become a lecturer in inorganic and physical chemistry in 1913.

He was conscripted into the army in August 1914 and spent the period 1916–18 as a "gas-protection officer" at the Physikalisch-chemisches Institut of Berlin University. Volmer's "research for the war effort" was of marginal importance, in contrast to his contact with Walther Nernst (1864–1941), whose thoughts on physics influenced Volmer and strengthened his drive to seek out the fundamental principles in all scientific problems. At Nernst's institute he also worked with the physicist Otto Stern (1888–1969). Their Stern-Volmer equation for the intensity of fluorescent light is still valid today. He also developed and patented the mercury vapour jet pump. Volmer distinguished himself by having a sharp eye for the technical application or commercial potential of his research.

After a short period as the director of the scientific laboratory of the Auer-Gesellschaft, he took a post at Hamburg University in 1920, but then changed two years later to become professor and director of the Institut für physikalische Chemie und Elektrochemie (Institute for Physical Chemistry and Electrochemistry) at the TH Berlin. His main fields of research were phase formation and crystal growth as well as power surg-

der Wissenschaften begannen und 1943/44 mit einem Dienststrafverfahren endeten. Die Situation am Institut verschlechterte sich, Mitarbeiter emigrierten oder wurden politisch verfolgt. Volmer resignierte und zog sich aus dem Institut zurück. Dennoch erschien 1939 sein Hauptwerk „Kinetik der Phasenbildung".

Nach Kriegsende sah er in Deutschland keine Arbeitsmöglichkeiten mehr und verpflichtete sich, seine Arbeiten in der Sowjetunion fortzusetzen. Dort widmete er sich der Deuteriumgewinnung und der Brennstoffaufbereitung, war an der eigentlichen Reaktorentwicklung jedoch nicht beteiligt. Als Volmer im Mai 1955 zurückkehrte, übernahm er, nach der Ernennung zum Professor für Physikalische Chemie an der Humboldt-Universität, die Präsidentschaft der Deutschen Akademie der Wissenschaften zu Berlin. Seine Amtszeit wurde durch die Umstrukturierung und Neugründung von Forschungseinrichtungen sowie durch die Auseinandersetzungen zur Verhinderung eines Atomkriegs geprägt. Wegen Krankheit legte Volmer im Oktober 1958 sein Amt nieder und blieb bis zu seinem Tode am 3.6.1965 Vizepräsident. Ihm zu Ehren wurde das Institut für physikalische Chemie der TU Berlin 1952 in „Max-Volmer-Institut" benannt, und er wurde 1955 von der TU Berlin zum Ehrendoktor ernannt.

Lit.: Blumtritt, Oskar : Max Volmer 1885–1965 : eine Biographie. – Berlin : TUB, 1985.

[M. E.]

es in electrode processes. After the rise of the Nazis in 1933 he was subject to various repressive measures, starting with the denial of his election to the Prussian Academy of Sciences and ending with disciplinary proceedings being brought up against him in 1933–1934. The situation at the institute worsened and employees emigrated or suffered political persecution. Volmer resigned and withdrew from the institute. Despite this, his main work "Kinetik der Phasenbildung", on the kinetics of phase formation, appeared in 1939.

After the war he saw no opportunities for work in Germany and decided to continue his research in the Soviet Union. There he devoted himself to the extraction of deuterium and to fuel treatment processes, but he did not actually take part in the development of nuclear reactors. When he returned to Germany in 1955, he was named professor of physical chemistry at the Humboldt University and also president of the German Academy of Sciences in Berlin. His tenure was marked by restructuring and the foundation of new research facilities as well as by the dispute over the prevention of nuclear war. Volmer resigned from his post in October 1958 because of illness but he remained vice-president until his death on 3 June 1965. The Institut für physikalische Chemie at the TU Berlin was renamed the Max-Volmer-Institut in his honour in 1952 and the university awarded him an honorary doctorate in 1955.

Lit.: Blumtritt, Oskar : Max Volmer 1885–1965 : eine Biographie. – Berlin : TUB, 1985.

[M. E.]

Peter Wapnewski (*1922)

Der wahnhafte Nationalismus war endgültig verflogen, deshalb war es unvermeidlich geworden, dass insbesondere eine Wissenschaft wie die Deutsche Philologie wieder einmal ihr Selbstverständnis zu überdenken hatte. „Ihre Wandlungsfähigkeit kann dieser Wissenschaft Bestätigung, Erneuerung oder Verlust der Substanz bringen", schrieb *Peter Wapnewski* dazu vor über 40 Jahren; und Philologen wie ihm verdanken wir es, dass die Substanz keinesfalls verloren gegangen ist, vielmehr neu entdeckt und die Deutsche Philologie folglich erneuert worden ist.

Peter Wapnewski wurde am 7. September 1922 in Kiel geboren und studierte von 1943 bis 1945 in Berlin, Freiburg, Jena in einer „Cavalierstour durch Fakultäten und Fächer" Kulturwissenschaften und von 1945 bis 1949 in Hamburg Ältere Deutsche Philologie und Klassische Archäologie; 1949 promovierte er dort mit der Schrift *Die Übersetzungen mittelhochdeutscher Lyrik im 19. und 20. Jahrhundert* und habilitierte sich im Jahre 1954 an der Universität Heidelberg mit einer Arbeit über Wolframs *Parzival*. Von 1956 bis 1958 lehrte er in Tübingen, 1959 wurde er als Professor der Deutschen Philologie nach Heidelberg berufen und 1966 an die Freie Universität Berlin. 1969 wechselte er nach Karlsruhe – „Berlin drohe Provinz zu werden", meinte er damals –, 1980 wurde er Gründungsrektor des Wissenschaftskollegs zu Berlin und verhalf Berlin damit aus dessen provinziellem Dasein heraus; 1982 berief ihn schließlich die TU Berlin, wo er bis zu seiner Emeritierung im Jahre 1990 lehrte.

„Uns ist in alten mæren wunders vil geseit
von helden lobebæren, von grôzer arebeit."

Wer mit diesen Versen anhebt, der erzählt uns mit dem *Nibelungenlied* in der Tat viel Bewundernswertes von ruhmreichen Helden und großer Mühsal, ja fast zu viel, wenn wir an den Nibelungenmythos denken, welcher der Verruf des *Nibelungenliedes* ist; und dadurch wurde dem

The insanity of National Socialism had finally subsided and so it was inevitable that a discipline like German philology had to particularly reappraise its view of itself. "Its adaptability can give the discipline validation, rejuvenation or loss of substance" wrote *Peter Wapnewski* on the subject 40 years ago, and we have to be grateful to philologists like him that the substance was not lost at all but was indeed re-discovered and German philology consequently revived.

Peter Wapnewski was born in Kiel on 7 September 1922 and studied arts and humanities in Berlin, Freiburg and Jena on a "Chivalrous tour through faculties and subjects" between 1943 and 1945. He studied Old German Philology and Classical Archeology at Hamburg from 1945 to 1949 and was awarded a doctorate in 1949 for his paper *Die Übersetzungen mittelhochdeutscher Lyrik im 19. und 20. Jahrhundert* (The translations of Middle High German lyric poetry in the 19th and 20th centuries) before qualifying as a lecturer in 1954 at the University of Heidelberg with a work on Wolfram's *Parzival*. He lectured at Tübingen from 1956 to 1958, and was appointed Professor of German Philology in Heidelberg in 1959 and at the Free University of Berlin in 1966. In 1969 he moved to Karlsruhe ("Berlin was threatening to become provincial" he said at the time) and in 1980 became the founding principal of Berlin's science college, so helping Berlin out of its provincial existence. Finally, he was called to the Technische Universität Berlin in 1982, where he taught until being granted emeritus status in 1990.

> *»Uns ist in alten mœren wunders vil geseit*
> *von helden lobebœren, von grôzer arebeit.«*
> (The myths of old recall many wonders,
> of gallant heroes and of supreme efforts.)

Anyone who begins quoting these verses actually tells us a great deal that is admirable about the *Nibelungenlied* with its glorious heroes

Literaturliebenden wie bei kaum einem anderen mittelalterlichen Werk
die Sicht auf die Schönheit und das Rätselhafte dieser Dichtung verstellt.
Doch Aufsätze wie Peter Wapnewskis Interpretation der „zartesten Szene
des ganzen Epos" – Hagen erbittet Rüdigers Schild – vermögen einem den
Reiz des *Nibelungenliedes* von neuem zu erschließen, und gleiches gilt
hinsichtlich seiner übrigen Untersuchungen über mittelalterliche Dich-
tungen, ob *Rolandslied, Tristan, Parzival, Minnesang* oder das *Tagelied*
Heinrichs von Morungen. Mit Recht heißt es in einer Zeitschrift für Litera-
turgeschichte über ihn, dass er „Modellstudien" geliefert habe; dies trifft
insbesondere auf die Form zu, denn Peter Wapnewski gehört zu jenen
Wissenschaftlern, die nicht wissen, weshalb sie unverständlich schreiben
sollen, wenn sie doch verstanden werden möchten – und die folglich in
klarer Ausdrucksweise verständlich schreiben. Schon seine wegweisende
Habilitationschrift fällt durch ihre Lesbarkeit aus dem Rahmen – auch
heute noch. Überdies zählt er zu jenen Germanisten, die einen eleganten
Sprachstil im Deutschen keinesfalls als ein Verfallszeichen der Deutschen
Philologie deuten. Deshalb seien Peter Wapnewskis *Zuschreibungen* emp-
fohlen, die neben Beiträgen über das Mittelalter auch welche über die
Neuere Literatur und über Musik enthalten und damit einen lesenswerten
Querschnitt seines gesamten Schaffens bilden.

------

*Lit.: Wolframs Parzival. Studien zur Religiosität und Form. Heidelberg 1955. – Deutsche
Literatur des Mittelalters. Ein Abriß von den Anfängen bis zum Ende der Blütezeit. 2.
Auflage Göttingen 1971. – Zuschreibungen. Gesammelte Schriften. Hildesheim und
Zürich 1994.*

[F. H.]

and great tribulations, perhaps almost too much if we think about the myth of the *Nibelungen*, which discredits the *Nibelungenlied*. As a result, literature lovers have a more greatly obscured conception of the beauty and the mysteriousness of the poetry than with any other medieval work. Yet essays such as Peter Wapnewski's interpretation of the "tenderest scene of the whole epic poem" (Hagen asking for Rüdiger's shield) give us a new approach for enjoying the *Nibelungenlied's* appeal, and the same applies with regard to his other examinations of medieval poetry, such as the *Rolandslied, Tristan, Parzival*, the *Minnesang* of the troubadours, or the *Tagelied* of Heinrich von Morungen. Quite correctly, one periodical for the history of literature wrote of him that he had delivered "model studies" and this applies in particular to the studies' structure, for Peter Wapnewski belongs to those academics who do not know why they should write incomprehensibly when they really want to be understood – and who consequently express themselves clearly. Even his pioneering postdoctoral thesis was exemplary for its understandability – and it still is. Moreover, he counts as one of those Germanists who by no means interpret an elegant style of language as a sign of decay in German philology. Thus Peter Wapnewski's *Zuschreibungen* (Attributions) are to be recommended. Alongside contributions on the Middle Ages, the work also contains texts on more modern literature and music, so providing a readable cross-section of his entire creative work.

Lit.: Wolframs Parzival. Studien zur Religiosität und Form. Heidelberg 1955. – Deutsche Literatur des Mittelalters. Ein Abriß von den Anfängen bis zum Ende der Blütezeit. 2nd edition, Göttingen 1971. – Zuschreibungen, Gesammelte Schriften. Hildesheim and Zurich 1994.

[F. H.]

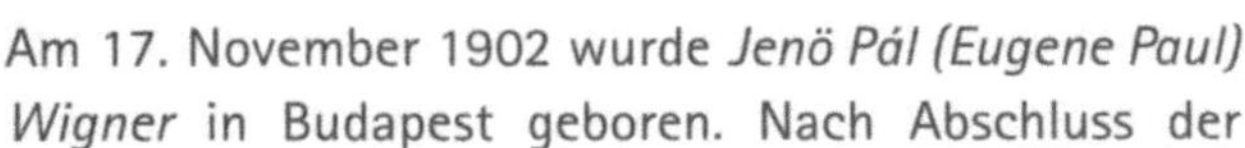

Am 17. November 1902 wurde *Jenö Pál (Eugene Paul) Wigner* in Budapest geboren. Nach Abschluss der Schule 1920 begann er ein Chemiestudium an der Technischen Hochschule in Budapest und wechselte 1921 an die TH Berlin. Das Studium schloss er 1924 mit einer Diplomarbeit ab und promovierte im Jahr darauf. Anschließend kehrte er nach Budapest zurück, um eine Stelle in einer Lederfabrik seines Vaters anzutreten.

1926 erhielt Wigner eine Assistentenstelle am Kaiser-Wilhelm-Institut für physikalische Chemie und Elektrochemie in Berlin. Noch im selben Jahr wurde Wigner Assistent von Richard Becker (1887–1955), der gerade die neu eingerichtete Professur für theoretische Physik an der TH Berlin übernommen hatte. An der TH begann Wigner seine Arbeiten zur Anwendung der Gruppentheorie auf die Quantenmechanik.

1927 ging Wigner für ein Jahr nach Göttingen. Hier traf er seinen Schulfreund John von Neumann (1903–1957) wieder, mit dem er im folgenden Jahr mehrere Artikel publizierte. Außerdem veröffentlichte er mit Pascual Jordan (1902–1980) eine Arbeit zum Pauliprinzip, sowie mit Viktor Weißkopf (1908–2002) zwei einflussreiche Artikel über die Breite von Spektrallinien. Schließlich entstand in Göttingen eine Arbeit, in der Wigner die Bedeutung der Symmetrien für die Quantenmechanik herausarbeitete und den Erhaltungssatz für die Parität formulierte.

1928 kehrte Wigner an die TH Berlin zurück und habilitierte sich. 1930 erhielt er eine befristete Halbzeit-Professur an der Universität Princeton (USA) und wurde gleichzeitig zum außerordentlichen Professor an der TH Berlin ernannt, so pendelte Wigner die nächsten Jahre zwischen Princeton und Berlin. Mit der Machtübernahme der Nationalsozialisten wurde seine Stelle an der TH jedoch sofort gestrichen, da Wigner Jude war. 1936 erhielt Wigner eine Professur an der University of Wiscon-

Jenö Pál (Eugene Paul) Wigner was born in Budapest, Hungary on 17 November 1902. He finished school in 1920 and started to study chemistry at the TH Budapest, moving to the TH Berlin the following year. He graduated in 1924 and took his doctorate in 1925. He then returned to Budapest to work at his father's leather factory.

In 1926, Wigner got a job as an assistant at the Kaiser-Wilhelm Institute for Physical Chemistry and Electrochemistry in Berlin. In the same year, he became assistant to Richard Becker (1887–1955), who had just taken on the newly-created professorship of Theoretical Physics at the TH Berlin. It was there that Wigner embarked on his research into the application of group theory to quantum mechanics.

In 1927 Wigner went to Göttingen for a year. There he met his old school friend John von Neumann (1903–1957) and in the following year the two published several articles. Wigner also co-authored a paper on the Pauli Exclusion Principle with Pascual Jordan (1902–1980) and he published two influential articles on spectral-line width with Viktor Weißkopf (1908–2002). Finally, he produced a paper in Göttingen, in which he worked out the significance of symmetries for quantum mechanics and where he formulated the parity conservation law.

In 1928 Wigner returned to the TH Berlin and became professor. In 1930 he became a temporary part-time professor in Princeton University in the United States, while at the same time he became professor with special status in Berlin. This meant that Wigner shuttled back and forth between Berlin and Princeton over the next several years. With the rise to power of the Nazis in Germany during the 1930s, Wigner's position at the TH Berlin was taken from him on the grounds that he was Jewish. In 1936, he took up a professorship at the University of Wisconsin and he

sin, im Jahr 1937 nahm er die amerikanische Staatsbürgerschaft an, und 1938 kehrte er auf eine Professur an die Princeton University zurück.

In Princeton wandte sich Wigner der quantenmechanischen Beschreibung von Festkörpern zu und arbeitete auf dem Gebiet der Kernphysik. Er lieferte wesentliche Beiträge zur Theorie der starken Kraft und veröffentlichte 1936 zusammen mit Gregory Breit eine Arbeit zum Neutroneneinfang in schweren Kernen. 1942 ging Wigner an die Universität Chicago, wo er im Rahmen des Manhattan-Projekts an der Entwicklung des Hanford-Reaktors zur Erzeugung von spaltbarem Plutonium arbeitete.

Nach dem zweiten Weltkrieg kehrte Wigner nach Princeton zurück, wo er weiter an Problemen der Kernreaktionen arbeitete sowie auf dem Gebiet der relativistischen Quantenmechanik. 1946/47 war er Direktor des Clinton Laboratory in Oak Ridge (Tennessee), wo Radioisotope für medizinische und andere zivile Anwendungen hergestellt wurden. Auch danach setzte er sich intensiv für die zivile Nutzung der Kernenergie ein und entwickelte weitreichende Aktivitäten auf dem Gebiet der Zivilverteidigung gegen einen Nuklearkrieg.

Wigner hat für seine Arbeit unzählige Preise und Auszeichnungen erhalten, als wichtigste sicherlich den Nobelpreis 1963. Mehr als 20 amerikanische und europäische Universitäten haben Wigner einen Ehrendoktor verliehen, darunter die Technische Universität Berlin im Jahr 1966.

Lit.: Jagdish Mehra: Eugene Paul Wigner: A Biographical Sketch, in The Collected Works of Eugene Paul Wigner, Vol. I, Berlin; Heidelberg; New York: Springer 1993

[J. Z.]

became an American citizen in 1937. The following year, he returned to Princeton.

At Princeton, Wigner devoted his efforts to producing a quantum mechanical description of solid objects and he also worked in the field of nuclear physics. He made considerable contributions to the theory of strong force and, together with Gregory Breit, published a paper on the capture of slow neutrons. In 1942, he went to the University of Chicago, where he worked on the Hanford Reactor for the production of fissionable plutonium as part of the Manhattan Project.

After World War II, Wigner returned to Princeton once again, where he continued his work on nuclear reactions as well as on relativistic quantum mechanics. He was director of the Clinton Library in Oak Ridge, Tennessee during 1946 and 1947, where radioisotopes were developed for medical and other civilian purposes. After that he continued to work intensively on civilian applications of nuclear technology and he developed wide-ranging projects in the area of civil defence for the event of nuclear war.

Wigner received numerous prizes and honours during his career, the most important among them being the Nobel Prize for Physics in 1963. More than 20 European and American universities conferred honorary doctorates on Wigner, one of them being the TU Berlin in 1966.

Lit.: Jagdish Mehra: Eugene Paul Wigner: A Biographical Sketch, in The Collected Works of Eugene Paul Wigner, Vol. I, Berlin; Heidelberg; New York: Springer 1993

[J. Z.]

Otto Nikolaus Witt (1853–1915)

Als sich *Otto Nikolaus Witt* 32-jährig an der TH Berlin für „Bleicherei, Färberei und Zeugdruck" habilitierte, konnte er nicht nur Promotion, 10 Jahre Industrietätigkeit und Lehrerfahrung vorweisen, sondern hatte auch den Azo-Farbstoff Chrysoidin erfunden und eine Farbstofftheorie aufgestellt, die – modifiziert – bis in unsere Tage gültig ist.

Witt wurde am 31.3.1853 in St. Petersburg als Sohn eines deutschen Pharmazeuten und einer Deutsch-Russin geboren. 1866 zog die Familie nach Zürich, wo Witt die Industrieschule (etwa Oberrealschule) absolvierte. Ab 1871 studierte er am Eidgenössischen Polytechnikum (heute ETH) Zürich Chemie und ging 1873 als Analytiker an eine Eisenhütte nach Duisburg. Seit 1874 Colorist der Kattundruckerei Schießer in Zürich, wurde er 1875 am Polytechnikum promoviert. Es folgten vier Jahre als Erster Chemiker in einer englischen Anilinfarbenfabrik, nach einem weiteren Jahr in Frankfurt/M. wurde er 1880 Lehrer an der Chemieschule in Mühlhausen/Elsass. Ab 1882 war er wissenschaftlicher Leiter in der Azofarben-Industrie in Mannheim. 1885 kam Witt nach Berlin, wo er, 1886 habilitiert, an der TH als Privatdozent lehrte und forschte und 1894 Ordinarius für Chemische Technologie wurde.

1876 hatte Witt mit der Erfindung des Chrysoidins die Synthese einer ganzen Palette gelber bis purpurner Azo-Farbstoffe durch sein Verfahren von Diazotieren und Kuppeln ermöglicht. Das Chrysoidin war keine zufällige Erfindung wie die ersten synthetischen Teerfarben, sondern beruhte auf theoretischen Überlegungen Witts, wie eine Lücke zwischen zwei schon bekannten Farbstoffen zu schließen sei. Im gleichen Jahr stellte er die Theorie auf, wonach Chromogene (meist aromatische Gruppen) durch Chromophore (Azo-, Nitro-, chinoide Gruppen) farbig und durch Auxochrome (Salzbildner, heute: Substituenten mit freien Elektronenpaaren) an das Gewebe gebunden werden. Der Farbstoffcharakter einer Substanz konnte damit erstmals sicher vorausgesagt werden.

When, at the age of 32, *Otto Nikolaus Witt* became
professor of "bleaching, dyeing and textile printing"
at the TH Berlin, his credentials included not only a doctorate and ten
years' experience in industry and teaching; he had also invented the azo
dye chrysodine and had proposed a theory of dyes, a modified version of
which is still in use today.

Witt was born on 31 March 1853 in Saint Petersburg, Russia. His
father was a pharmacist and his mother came from a Russo-German
family. In 1866 the family moved to Zurich, Switzerland, where Otto
attended the Industrieschule. From 1871 he studied Chemistry at the
Eidgenössische Technische Hochschule (now the Swiss Federal Institute
of Technology) and in 1873 went to the German town of Duisburg as an
analyst at an ironworks. From 1874 he was a colourist at the Schießer
cotton printing factory in Zurich, and in 1875 he took his doctorate at the
Eidgenössische Technische Hochschule. There then followed four years as
chief chemist in an English factory making aniline dyes. After a further
year in Frankfurt, in 1880 he became a teacher at the chemistry school in
Mühlhausen (Mulhouse, France). From 1882 he was head scientist at the
Azofarben-Industrie in Mannheim. In 1885 Witt came to Berlin, where he
became a professor in 1886, taught and researched at the TH as a private
teacher and in 1894 became the professor of chemical technology.

With his invention of chrysoidine in 1876, Witt had made possible
the synthesis of an entire palette of azo dyes from yellow to purple using
his technique of diazotation and coupling. Unlike the first synthetic coal
tar dyes, chrysoidine was not a chance invention, rather it was based on
Witt's theoretical conclusion that there was a gap between two already
known dyestuff-types. Also in 1876, he proposed the theory that chro-
mogens (aromatic structures) gain their colour by means of chromophores
(the group of chemicals containing the azo, quinoid and nitro- groups),
and they can be attached to textiles by means of auxochromes (the salt-

An der TH beschäftigte sich Witt zudem mit Silikatchemie (Keramik, Glas) und legte eine umfassende, auch missratene Stücke umfassende Lehrsammlung an. Nicht Spezialistentum, sondern die Anwendung der Chemie als Ganzes war sein Ausbildungsziel, wobei er besonderes Gewicht darauf legte, das wissenschaftliche Denken dem Wesen und den Erfordernissen der Technik anzupassen. Für eine wirklichkeitsnahe Lehre hielt er engen Kontakt zu chemischen Betrieben. Bei der Einweihung des von ihm angeregten und in der Einrichtung geplanten neuen technisch-chemischen Instituts 1905 wurde sein didaktisches Geschick offenbar – er setzte sogar schon Filme als Unterrichtsmittel ein.

Als gefragter Gutachter mit dem Patentwesen vertraut, präzisierte er Begriffe wie „Erfindungsäquivalent" für Industrieerfordernisse. Seine über 110 Arbeiten und die Herausgabe der Zeitschrift „Die chemische Industrie" bereicherten die Wissenschaft. Mit Berichten von den Weltausstellungen 1893 und 1900, vor allem aber mit seiner Zeitschrift „Prometheus" trug er wesentlich zur Popularisierung der Naturwissenschaften bei. Er wurde mit zahlreichen Orden und Auszeichnungen hoch geehrt. Zudem war Witt ein anerkannter Diatomeenforscher, züchtete Orchideen und schrieb Gedichte. Nach einem Vortrag in der Deutschen Chemischen Gesellschaft am 23.3.1915 erlag er einem Schlaganfall.

Lit.: Emilio Noelting: Otto Nikolaus Witt, 1853–1915, in: Berichte der Deutschen Chemischen Gesellschaft 49 (1916), S. 1751–1832; Jean D'Ans: Zum 100. Geburtstage von Otto Nikolaus Witt, in: Chemiker-Zeitung 77 (1953), S. 279–281.

[B. E.]

forming group of chemicals, later recognised as electron-donating substances). This meant that the dyeing characteristics of a given substance could be predicted for the first time.

At the TH Berlin, Witt also researched silicates (ceramic, glass) and he laid out an extensive, if flawed, collection of specimens for teaching. His educational goal was not specialization, rather the application of chemistry as a whole. To this end he laid a special emphasis on combining scientific thought with the nature and the prerequisites of technology. In order to keep his teaching grounded in real practice, he retained close contact with the chemical industry. When a new institute for technology and chemistry was set up in 1905, at Witt's instigation and with his planning behind it, his didactical contribution was clear to see – he even included films as teaching materials.

As a sought-after expert relating to patents, he helped define legal-scientific terms which were required by industry. He made a great contribution to science with his more than 110 papers and his magazine "Die chemische Industrie". His reports, mainly in his magazine "Prometheus", on the World Exhibitions of 1893 and 1900, made significant contributions to the popularization of the natural sciences. He was honoured with numerous awards. Witt was also a recognised expert on diatoms (unicellular organisms), he bred orchids and wrote poetry. He died of a stroke on 23 March 1915 after giving a speech to the Deutsche Chemische Gesellschaft (German Chemistry Society).

Lit.: Emilio Noelting: Otto Nikolaus Witt, 1853–1915, in: Berichte der Deutschen Chemischen Gesellschaft 49 (1916), p. 1751–1832; Jean D'Ans: Zum 100. Geburtstage von Otto Nikolaus Witt, in: Chemiker-Zeitung 77 (1953), p. 279–281.

[B. E.]

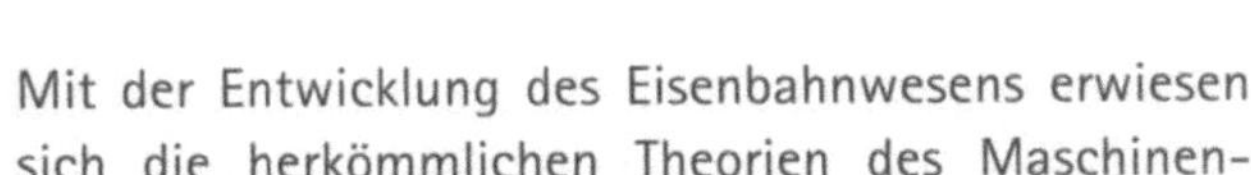

August Wöhler (1819–1914)

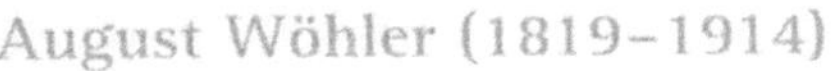

Mit der Entwicklung des Eisenbahnwesens erwiesen sich die herkömmlichen Theorien des Maschinenbaus als unzureichend. So entstanden als Mittler zwischen Konsument und Produzent die Materialforschung und -prüfung, zu deren Pionieren *August Wöhler* zählt.

Wöhler wurde am 22.6.1819 in Soltau als Sohn eines Lehrers geboren. Durch seine mathematische Begabung aufgefallen, kam er 1835 auf die Höhere Gewerbeschule (später TH) Hannover mit einem jährlichen Stipendium von 100 Talern. Die Auflage, am Studienende ganztags in einer Werkstatt zu arbeiten, legte wohl die Basis zur späteren Entwicklung der so einfachen wie effizienten Prüfgeräte.

Mit vorzüglichem Zeugnis verließ Wöhler die Gewerbeschule, erhielt ein Reisestipendium von 100 Talern, das ihn bewog, nach Berlin zu gehen. 1841–43 arbeitete er bei August Borsig (1804–1854), danach bei der Kgl. Hannoverschen Eisenbahn und wechselte 1847 als Obermaschinenmeister zur Niederschlesisch-Märkischen Eisenbahn in Frankfurt/Oder. Nach deren Übernahme durch Preußen blieb Wöhler, 1854 zum Kgl. Preuß. Obermaschinenmeister ernannt, bis 1869 im Staatsdienst und trat 1869 als Direktor in die Norddeutsche Aktiengesellschaft für Eisenbahnbedarf in Berlin ein. 1874 wurde er als Eisenbahndirektor und Mitglied der Generaldirektion der Reichseisenbahnen nach Straßburg berufen und trat 1889 in den Ruhestand. Wöhler starb am 21.3.1914 in Hannover.

In einer Ministerialkommission war er seit 1852 maßgeblich an der Entwicklung von Dauerbelastungsmessungen von Eisenbahnteilen zur Festlegung allgemeiner Konstruktionsprinzipien im Lokomotiv- und Waggonbau, an der Gestaltung von Schienenprofilen, Achsen, Radreifen und Bremsen, aber auch an Brückenkonstruktionen beteiligt. Als Mitglied des Technischen Ausschusses im Verein Deutscher Eisenbahnverwaltungen trat er für die Einführung einer staatlich anerkannten Klassifikation von

With the development of the railways, the conventional theoretical basis of engineering was shown to be lacking. The disciplines of materials research and materials testing emerged to bridge the divide between the user and the producer, and one of the pioneers in the field was *August Wöhler.*

Wöhler was born on 22 June 1819 in Soltau, the son of a teacher. He had a striking facility in mathematics, and entered the Höhere Gewerbeschule (later the TH) in Hanover in 1835 with an annual scholarship of 100 thalers. The students were required to work in the workshop all day, and this laid the basis for his later development of simple and efficient testing instruments.

Wöhler left the Gewerbeschule with excellent marks and he won a travelling scholarship of 100 thalers, which enabled him to go to Berlin. He worked with August Borsig (1804–1854) in the period 1841–43, then with the Kgl. Hannoversche Eisenbahn and then in 1847 he became Obermaschinenmeister in the Niederschlesisch-Märkische railway in Frankfurt/Oder. When these were taken over by Prussia, Wöhler stayed on, becoming Royal Prussian Obermaschinenmeister in 1854, and he remained working for the state until 1869, when he became director of the Norddeutsche Aktiengesellschaft für Eisenbahnbedarf in Berlin. In 1874 he moved to Straßburg (Strasbourg, France) to take up the post of railway director and member of the Generaldirektion der Reichseisenbahnen. He retired in 1889 and died in Hanover on 21 March 1914.

From 1852 Wöhler was a member of a ministerial commission, and as such was instrumental in several advances: the development of ways of measuring continuous fatigue in order to determine general construction principles for the building of locomotives and wagons, the design of rail profiles, axes, tyres and brakes, as well as bridge construction. As a member of the Technical Committee of the Verein Deutscher Eisenbahn-

Stahl und Eisen ein und forderte die Gründung von Materialprüfungs- und Versuchseinrichtungen. Er beteiligte sich weiter an der theoretischen und praktischen Ausbildung von Prüfverfahren, überließ jedoch die praktische Durchführung seinem Assistenten. Mit der Konstruktion von Apparaturen für Dauerbelastungsversuche betrat Wöhler methodisches und versuchstechnisches Neuland und förderte die Veränderung der Denkweisen in den technischen Disziplinen maßgeblich.

Seine Frankfurter Versuche waren in Fachkreisen zunächst auf wenig Resonanz gestoßen, doch wurde 1870 der Berliner Gewerbeakademie die Weiterführung der Wöhlerschen Versuche mit den originalen Apparaturen angeordnet. Das war der Geburtstag der Mechanisch-technischen Versuchsanstalt, aus der das Preußische Materialprüfungsamt und schließlich die Bundesanstalt für Materialprüfung (BAM) – jetzt Bundesanstalt für Materialforschung und –prüfung – hervorging.

Wöhlers Interessen blieben auf das Eisenbahnwesen beschränkt, doch dauerte es nicht allzu lange, bis sich Materialforschung und –prüfung auf alle Bereiche der Technik ausdehnten. 1901 gehörte Wöhler zu den ersten, die von der gerade mit dem Promotionsrecht versehenen TH Berlin den Ehrendoktortitel erhielten.

Lit.: R. Blaum: August Wöhler, in: Beiträge zur Geschichte der Technik und Industrie 8 (1918), S. 34–55; Walter Ruske: August Wöhler (1819–1914); Zur 150 Wiederkehr seines Geburtstages, in: Materialprüfung 11 (1969), S. 181–188; Walter Ruske: 100 Jahre Materialprüfung in Berlin, Berlin 1971.

[M. E.]

verwaltungen, he was in favour of introducing state-approved classifications for steel and iron and he also called for the foundation of facilities for materials testing and testing equipment. He took part in teaching the theoretical and practical training of testing methods, but he left the practicalities of carrying them out to his assistant. With the construction of equipment for testing fatigue, Wöhler broke new ground in terms of methodology and testing technology, and it had a significant impact on modes of thought in the technical disciplines.

At first, his experiments at Frankfurt had little resonance in specialist circles, but in 1870 the Berliner Gewerbeakademie was instructed to continue Wöhler's experiments with the original equipment. This may be regarded as the birth of the Mechanisch-technische Versuchsanstalt, which in turn became the Preußische Materialprüfungsamt and eventually the Bundesanstalt für Materialprüfung – which today has become the Bundesanstalt für Materialforschung und –prüfung (Federal Institute for Materials Research and Testing).

Wöhler's field of interest remained restricted to railways, but it did not take long for materials research and materials testing to find their way into all aspects of technical activity. In 1901 Wöhler was one of the first people to be awarded an honorary doctorate from the TH Berlin, which had just been granted the right to confer degrees.

Lit.: R. Blaum: August Wöhler, in: Beiträge zur Geschichte der Technik und Industrie 8 (1918), p. 34–55; Walter Ruske: August Wöhler (1819–1914); Zur 150 Wiederkehr seines Geburtstages, in: Materialprüfung 11 (1969), p. 181–188; Walter Ruske: 100 Jahre Materialprüfung in Berlin, Berlin 1971.

[M. E.]

Karl Zander (*1923)

Karl Zander wurde am 10. April 1923 in Schwerin geboren. Nach dem Abitur 1941 wurde er zum Wehrdienst eingezogen und blieb bis Kriegsende im Einsatz bei der Heeres-Artillerie. Von 1946 bis 1951 studierte Zander Elektrotechnik an der TU Berlin und wurde anschließend Assistent am Lehrstuhl für Theoretische Elektrotechnik.

Als Entwicklungsingenieur bei der Firma AEG, den Posten hatte Zander 1956 angetreten, setzte er die damals neuen Fototransistoren für die Steuerung eines Walzwerkes ein, die erste europäische Entwicklung dieser Art. 1957 promovierte er an der TU und wurde 1958 Leiter der Abteilung Elektronik im neu gegründeten Institut für Kernforschung Berlin (später Hahn-Meitner-Institut - HMI). In der Folgezeit leitete er maßgeblich den Aufbau zahlreicher Großgeräte am HMI. Zander beschäftigte sich außerdem weiter mit Halbleiterelektronik, über die er bereits 1960 an der TU eine Vorlesung anbot. 1963 habilitierte er sich an der TU mit einer Arbeit über das Verhalten von Halbleiterbauelementen unter dem Einfluss ionisierender Strahlung. Durch diese Forschung begründete er eine langfristige Arbeitsrichtung des HMI, Strahlungsrisiken für Satelliten in Modellrechnungen und in Simulations-Experimenten zu erfassen und strahlenresistente Bauelemente zu entwickeln. Höhepunkte waren Projektbegleitungen für die ersten europäischen Weltraum-Satelliten, aber auch für die deutschen Experimente bei NASA-Missionen zum Jupiter. 1966 wurde Zander in die Kommission für Weltraumforschung des Bundesministeriums für Forschung und Technik berufen.

Während eines einjährigen Forschungsaufenthalts am MIT 1966/67 befasste sich Zander mit der computergestützten Entwicklung (CAD) von Schaltungen. 1969 wurde Zander zum Ordinarius für Elektronik an die TU Berlin und gleichzeitig zum Direktor des Sektors Elektronik im HMI berufen. Durch die Vernetzung von TU- und HMI-Aktivitäten entwickelte sich in seinem Umkreis die seinerzeit modernste Forschung und Universitäts-

As a development engineer at AEG, where Zander began work in 1956, he used the (then) new phototransistors to control a rolling mill, the first development of its kind in Europe. In 1957 he was awarded a doctorate at the TU Berlin and in 1958 became head of the Department of Electronics at the re-established Berlin Institute for Nuclear Research (later known as the Hahn-Meitner-Institut or "HMI"). In the following years he was responsible for the assembly of numerous major items of equipment at the HMI. Zander was also involved in semiconductor electronics, a subject on which he had already given one lecture at the TU Berlin in 1960. In 1963 he qualified as a university lecturer at the TU Berlin with a work on the behaviour of semiconductor components under the influence of ionizing radiation. Through this research he set up a long-term HMI project to record the risks to satellites from radiation using calculations and simulations, and to develop radiation-resistant components. Highlights included projects connected with the first European space satellites and above all, for the German experiments on the NASA missions to Jupiter. In 1966 Zander was appointed a member of the Commission for Space Research of the Federal Ministry for Research and Technology.

Zander also conducted research into the computer-assisted design (CAD) of electrical circuitry during a year-long research stay at the MIT (1966/67). In 1969, he was appointed full professor of electronics at the TU Berlin and, at the same time, director of the HMI's electronics sector. Through this interlinking of TU and HMI activities, he played a cen-

ausbildung der Elektronik in Deutschland. Im gleichen Jahr initiierte er
die Gründung der Arbeitsgemeinschaft für Großforschungseinrichtungen,
der Vorläuferin der heutigen Helmholtz-Gesellschaft.

Besonders engagierte sich Zander für die Ausstattung der Hochschulen und Forschungsinstitute in Berlin mit modernen Rechnernetzen.
1974 begründete er das HMI-NET, den damals größten deutschen Rech-
nerverbund. Seine Empfehlungen führten zum Verbund der Universitäts-
rechenzentren mit dem Großrechenzentrum Berlin, woraus sich 1977 das
Berliner Rechnernetz für die Wissenschaft (BERNET) entwickelte. Auch
an der Entwicklung des Deutschen Forschungsnetzes (DFN) war Zander
maßgeblich beteiligt. Ebenso war er an der Planung und Entwicklung von
BERKOM, einem breitbandigen Glasfasernetz, sowie dem TU-internen
Breitbandnetz TUBKOM beteiligt. 1987 gründete Zander das Forschungszentrum für Offene Kommunikationssysteme der Gesellschaft für Mathematik und Datenverarbeitung, dessen Leitung er auch übernahm.

Für sein Lebenswerk wurde Zander 1985 mit dem Verdienstkreuz 1.
Klasse des Verdienstordens der Bundesrepublik Deutschland ausgezeichnet. 1991 wurde Zander emeritiert, und 2003 verlieh ihm die TU Berlin
die Ehrenmitgliedschaft.

[J. Z.]

tral role in the development of the most modern research and university training of electronics in Germany. The same year he was influential in the foundation of the working group for large-scale research facilities, the predecessor of the today's Helmholtz Association.

Zander was also deeply committed to equipping the universities and research institutes in Berlin with modern computer networks. In 1974 he founded HMI-NET, at the time the largest German computer network. His recommendations led to the establishment of the network of university computing centres with a major computer centre in Berlin, which (in 1977) became the Berlin Computer Network for Science (BERNET). Zander was also heavily involved in the development of the German Research Net (DFN). Similarly, he was active in the planning and development of BERKOM, a broadband glass fiber network, as well as TUBKOM – the TU's own internal broadband network. In 1987, Zander established the Research Center for Open Communication Systems at the Society for Mathematics and Data Processing, of which he became head.

In 1985, Zander was awarded the Distinguished Service Cross 1st Class of the Federal Republic of Germany in appreciation of his life's work and in 1991 he was conferred the status of emeritus professor and in 2003 honorary membership of the TU Berlin.

[J. Z.]

Der Computer: mein Lebenswerk. Wozu der vielen Worte, denn mit dem Titel seiner Autobiographie lässt sich *Konrad Zuses* Wirken kurz und bündig schildern.

Geboren wurde Konrad Zuse am 22. Juni 1910 in Berlin; in Ostpreußen und Brandenburg wuchs er auf und studierte seit 1928 an der TH Berlin – „nicht ohne Irr- und Seitenwege", wie er humorvoll bekannte. Denn nach einer Einheit aus Ingenieur und Künstler trachtend, verlegte er sich erst auf Maschinenbau, dann auf Architektur, sah seine Hoffnung jedoch jeweils nicht erfüllt und landete schließlich beim Bauwesen. Dieses Studium schloss er – nach einem Zwischenspiel als Reklamezeichner – 1935 zwar erfolgreich ab, aber nicht ohne seitdem eine herzliche Abneigung gegen statische Rechnungen zu hegen, womit die Bauingenieurstudenten aus Zuses Sicht gequält worden seien. Und das war ein Glück, denn so kam es bald darauf zu der Geburt des Computers aus dem Geiste öden Rechnens. Zuses Tätigkeit als Statiker in einem Flugzeugwerk ließ diese Abneigung nämlich so prächtig gedeihen, dass er nach nur einem Jahr seine Stelle kündigte, um sich künftig der Entwicklung einer Maschine zu widmen, welche diese Rechnungen automatisch durchführen sollte. Überlegungen hierzu waren bereits 1933 entstanden, nun ging er mit Hilfe einiger Freunde jene legendäre Arbeit an, der 1941 mit der Z3 die erste funktionsfähige programmierbare Rechenanlage entsprang.

1942 hatte Zuse mit dem Bau der Z4 begonnen, die er sogar durch die Kriegswirren retten konnte; doch die äußeren Umstände jener Jahre ließen seine praktische Arbeit vorübergehend zum Stillstand kommen, weshalb sich Zuse – weit mehr als nur ein genialer Bastler – zunächst mit der Entwicklung einer Programmiersprache, des Plankalküls, beschäftigte. Der Prophet galt zwar noch nichts im eigenen Lande, aber zumindest das Ausland war auf ihn aufmerksam geworden, und neben Besuchern aus England, Frankreich und den USA durfte er 1949 Eduard Stiefel von der ETH Zürich empfangen. Dieser hatte die Möglichkeiten der Z4 erkannt

Konrad Zuse was born in Berlin on 22 June 1910. He grew up in East Prussia and Brandenburg and began studying at the Technische Hochschule Berlin in 1928 – but not without taking some "wrong turnings and back roads" as he amusingly put it – because after spending a term striving to be an engineer and artist, he switched first to mechanical engineering and then to architecture. However, seeing his hopes unfulfilled, he ended up a student of building. Though he was able to successfully complete this course in 1935 (following an interlude as a graphic artist in the advertising trade) it was not without having developed a heartfelt aversion to static calculations, which Zuse considered to be a torture for students. And this dislike proved fortunate, for soon afterwards the computer was born, dreary arithmetic's offspring. During Zuse's activities as structural engineer in an aeroplane factory, his aversion became so great that he handed in his notice after just one year so he could devote himself to the development of a machine that would perform these calculations automatically. The first considerations in this direction were made as early as 1933, but then, with the help of a few friends, he began on a now-legendary effort that led to the birth of the "Z3", the first functioning, programmable computer system in 1941.

In 1942, Zuse had begun the construction of the Z4, a project he was able to continue even through the chaos of the Second World War, though the overall situation during those years forced his practical work to a standstill, which is why Zuse (who was much more than someone who was merely brilliant with his hands) first kept himself busy with the development of a programming language called *Plankalkül* (planned calculation). The prophet did fall on deaf ears in his own country but at least he had gained a reputation abroad, and alongside visitors from England,

und ließ 1950 die Anlage auf Ausleihbasis im Institut für angewandte Mathematik der ETH Zürich aufstellen. Von nun an ging es mit Zuses neugegründeter Firma bergauf, und nach einigen Jahren erwarb auch die TU Berlin mit der Z22 einen Rechner Zuses.

Der Versuch einer Zusammenarbeit mit der TH Berlin war für Zuse eher enttäuschend verlaufen, denn als 1938 ein Freund und er dort vor einem kleinen Kreise einen Vortrag über die elektronische Rechenmaschine hielten, ernteten sie Kopfschütteln; die Zuhörer sahen die zugegebenermaßen hohen Hürden, und Zuse gelang es nicht, sie dazu anzuspornen, die damit verbundenen hohen Herausforderungen bewältigen zu wollen. Immerhin erkannte ein Professor den Wert einer solchen Rechenmaschine, und an dessen Lehrstuhl erhielt Zuses Freund den Auftrag, ein Versuchsmodell eines Rechenwerks zu bauen. Nach dem Kriege war es insbesondere Wolfgang Haack zu verdanken, dass seit 1949/50 Kontakte zwischen der TU Berlin und Zuse geknüpft wurden; dieser hielt im Wintersemester 1953/54 eine Gastvorlesung, und vor allem ging 1958 die erwähnte Rechenmaschine Z22 in Betrieb, sodass nun auch die Universität, die Konrad Zuse auf Umwegen zur Erfindung des Computers inspiriert hat, endgültig in das neue Zeitalter eingetreten war.

Lit.: Über den Plankalkül als Mittel zur Formulierung schematisch-kombinativer Aufgaben. In: Archiv für Mathematik 1 (1948/49) – Der Plankalkül. St. Augustin 1972. – Der Computer: mein Lebenswerk. Berlin u. a. 1984.

[F. H.]

France and the USA, he was able to receive Eduard Stiefel from the ETH in Zurich in 1949. The latter had been able to recognize the opportunities the Z4 offered, and Stiefel had the unit assembled on loan at the ETH Zurich's Institute for Applied Mathematics in 1950. From then on, things got quickly better for Zuse's newly founded company, and a few years later the Technische Universität Berlin also acquired one of his "Z22" computers.

The attempt at cooperation with the Technische Hochschule Berlin had been rather disappointing for Zuse because, in 1938, when he and a friend had given a lecture on electronic computing devices before a small audience there, they had been received with much shaking of heads. Those attending saw the (admittedly large) obstacles ahead, and Zuse could not spur them on to meet the major challenges involved. Yet one professor was able to recognize the value of an electronic computing machine, and Zuse's friend was instructed to construct a test model of such a computing device at the professor's faculty in 1938. After the war, thanks were due in particular to Wolfgang Haack for establishing contacts between Zuse and the Technische Universität Berlin after 1949/50. Zuse gave a guest lecture during the 1953/54 winter term, and most importantly, the Z22 computing machine mentioned above started operations in 1958, so that the university which (in a roundabout way) had inspired Konrad Zuse to invent the computer could itself enter the new era at last.

Lit.: Über den Plankalkül als Mittel zur Formulierung schematisch-kombinativer Aufgaben. In: Archiv für Mathematik 1 (1948/49) – Der Plankalkül. St. Augustin 1972. – Der Computer: mein Lebenswerk. Berlin, etc. 1984.

[F. H.]

Personenregister/Name Index

Sachverzeichnis

Sachverzeichnis

Subject Index

Subject Index

Bildnachweis/List of Figures

Bildnachweis/List of Figures

Bildnachweis/List of Figures

Bildnachweis/List of Figures